Lecture Notes in Computer Science 3324

Commenced Publication in 1973
Founding and Former Series Editors:
Gerhard Goos, Juris Hartmanis, and Jan van Leeuwen

Ming-Chien Shan   Umeshwar Dayal
Meichun Hsu (Eds.)

Technologies
for E-Services

5th International Workshop, TES 2004
Toronto, Canada, August 29-30, 2004
Revised Selected Papers

 Springer

Volume Editors

Ming-Chien Shan
Umeshwar Dayal
Hewlett-Packard
1501 Page Mill road, Palo Alto
CA, 94304, USA
E-mail: {ming-chien.shan, umeshwar.dayal}@hp.com

Meichun Hsu
Commerce One Inc.
12717 Leander Dr. Los Altos Hills
CA 94022, USA
E-mail: mhsu@alum.mit.edu

Library of Congress Control Number: 2005921803

CR Subject Classification (1998): H.2, H.4, C.2, H.3, J.1, K.4.4, I.2.11

ISSN 0302-9743
ISBN 3-540-25049-2 Springer Berlin Heidelberg New York

Springer is a part of Springer Science+Business Media

springeronline.com

© Springer-Verlag Berlin Heidelberg 2005
Printed in Germany

Typesetting: Camera-ready by author, data conversion by Scientific Publishing Services, Chennai, India
Printed on acid-free paper   SPIN: 11398073   06/3142   5 4 3 2 1 0

Preface

The 2004 VLDB workshop on Technologies on E-Services (VLDB-TES 2004) was the fifth workshop in a series of annual workshops endorsed by the VLDB Conference. It served as a forum for the exchange of ideas, results and experiences in the area of e-services and e-business.

VLDB-TES 2004 took place in Toronto, Canada. It featured the presentation of 12 regular papers, focused on major aspects of e-business solutions. In addition, the workshop invited 2 industrial speakers to share their vision, insight and experience with the audience.

The workshop would not have been a success without help from so many people. Special thanks go to Fabio Casati, who organized the program agenda and the proceedings publication, and Chandra Srivastava, who served as the publicity chair. We also thank the members of the program committee and the additional reviewers for their thorough work, which greatly contributed to the quality of the final program.

We hope that the participants found the workshop interesting and stimulating, and we thank them for attending the workshop and for contributing to the discussions.

Toronto, Canada   Meichun Hsu
September 2004   Umeshwar Dayal
                                                                 Ming-Chien Shan

Organization

Conference Organization

Conference Chair: Ming-Chien Shan (Hewlett-Packard, USA)
Program Chairs: Meichun Hsu (CommerceOne, USA) and Umeshwar Dayal (Hewlett-Packard, USA)

Program Committee

Gustavo Alonso, ETH Zürich, Switzerland
Roger Barga, Microsoft, USA
Boualem Benatallah, University of New South Wales, Sydney, Australia
Christof Bornhoevd, SAP, USA
Christoph Bussler, DERI, Ireland
Fabio Casati, Hewlett-Packard, USA
Jen-Yao Chung, IBM, USA
Francisco Curbera, IBM, USA
Dimitrios Georgakopoulos, Telcordia, USA
Dean Jacob, BEA Systems, USA
Frank Leymann, IBM, Germany
Heiko Ludwig, IBM, USA
Mike Papazoglou, Tilburg University, the Netherlands
Barbara Pernici, Politecnico di Milano, Italy
Calton Pu, Georgia Tech, USA
Krithi Ramamritham, IIT Bombay, India
Farouk Toumani, LIMOS, France
Steve Vinoski, Iona, USA
Hartmut Vogler, SAP Research Lab, USA
Sanjiva Weeravaraana, IBM
Leon Zhao, University of Arizona, USA

Table of Contents

Robust Web Services via Interaction Contracts

David Lomet

Microsoft Research
lomet@microsoft.com

Abstract. Web services represent the latest effort of the information technology industry to provide a framework for cross enterprise automation. One principal characteristic of this framework is information hiding, where only the message protocol is visible to other services or client software. This environment makes it difficult to provide robust behavior for applications. Traditional transaction processing uses distributed transactions. But these involve inter-site dependencies that enterprises are likely to resist so as to preserve site autonomy. We propose a web services interaction contract (WSIC), a unilateral pledge by a web service. A WSIC avoids dependencies while enabling, but not requiring, an application to be written so that it can provide exactly once execution semantics, even in the presence of failures. And exactly once semantics is essential whenever commercial interactions involve money.

1 Introduction

1.1 Using the Web for Commerce

The importance of the web for commerce only continues to grow. But that growth has been a bit "cranky". Growth is impeded by the barriers of legacy systems, differing protocols, the lack of robustness, and a lack of an integrative framework. But things are changing, as industry is well aware of these difficulties. The current "horse" being ridden is that of "web services", which offers the start of a compelling story about how to break down these barriers and produce the flourishing of commerce on the web.

So, abstractly, what are web services? And why are web services the way to go? The web services effort places special emphasis on "information hiding", an idea that began with David Parnas over 30 years ago [10]. The idea is to minimize what applications need to know in order to interact meaningfully with the service. Web services exploit message protocols, using a common syntactic framework (XML with SOAP) but are otherwise opaque in terms of how they are implemented behind their service front. This makes it possible for diverse systems to interact, by hiding the "ugly details" behind the service interface.

There is a lot to the web services story (UDDI [9], WSDL [11], etc.) but the focus here will be on the robustness of web services and the applications that use them. Currently, web applications either are not very robust, or need hand crafting to be so. But for pervasive use of web services, we need to ensure both robustness and a standard, non-intrusive way of making it possible.

M.-C. Shan et al. (Eds.): TES 2004, LNCS 3324, pp. 1–14, 2005.

1.2 Making Web Services Robust

There is a long history of efforts to make distributed applications robust [6]. The goal of most of these efforts has been to reduce the work lost in a long running application due to a system crash. Commercial systems need to ensure that no committed work is lost. Most commercial efforts evolved in the transaction processing community, with TP monitors and distributed transactions [1, 7]. And such transaction processing efforts have been quite successful in the context of relatively closed systems. There are efforts to extend this style of support to the web services world (e.g. WS- Coordination [12]). However, TP has had much less impact on highly distributed, largely autonomous, and heterogeneous commercial systems.

Classic transaction processing relies heavily on distributed transactions. But even in the classic TP world, distributed transactions are not usually very "distributed". They are used mostly for partitioned databases and for stateless applications interacting with queue managers. It is rare for this technology to be used across enterprise boundaries, and is not widely used even across organizational boundaries within a single enterprise. One can hypothesize a number of reasons for this. Protocols are too tied to a particular implementation base. Or these protocols require too much compromise of autonomy. For example, who coordinates a transaction and who will block if there is a failure at the wrong time? But without distributed transactions, there are no TP-based solutions. The situation then reverts either to non-robust applications and systems or to roll-your-own special case solutions.

1.3 Interaction Contracts to the Rescue?

I want to make a case for "interaction contracts" [5] as the paradigm for robust web services. Interaction contracts enable applications to be persistent across system crashes, without distributed transactions. They enable applications to interact with "transactional services" with the application neither participating in the transaction nor perhaps even being aware that there is a transaction involved. Thus, atomicity can be encapsulated within an "opaque" web service, and supported by transactions whose distribution can be limited to the traditional TP domain, and whose participants could be databases, queue managers, etc.

It is important to emphasize here that atomicity within web services is still highly desirable, perhaps essential. One wants an e-commerce sale to be all or nothing: no money—no product; money—product. The in-between states are unacceptable. But whether this atomicity is provided by a single local transaction, a distributed transaction, a workflow system, or some other technology, should be immaterial at the level of the web service.

If the applications using web services are not participating in transactions, then how are they made robust? The answer here is that interaction contracts capture precisely what needs to be done for applications to be persistent, surviving system failures. These applications live outside of any transaction, and because they are persistent, they can robustly provide program logic for dealing with failures, which might have resulted from one or more transaction aborts or system crashes. This is important, as traditional TP with stateless applications had no convenient place to put program logic dealing with transaction failures.

There is more to robustness than persistence (the ability for an application to survive system crashes). Traditional TP can provide not just applications that survive system failures and have certain atomicity properties. It also provides solutions for other robustness attributes: availability, scalability, and load balancing. Fortunately, these attributes can also be provided with interaction contracts and persistent applications. We have discussed these attributes in papers describing our prototype Phoenix/App system [2, 3] that provides transparent support for persistent .NET applications [8]. So I will not discuss these here.

2 Interaction Contracts

2.1 Overview

Most of what is described in this section is based on recovery guarantees [4, 5] that provide exactly once execution. Interaction contracts are obligations for each party involved in sending or receiving a message. The intent of the obligations is to ensure that both sides of the interaction can agree, even in the presence of a system crash of one or both of the parties, as to what the interaction consisted of and whether it occurred. Each party to the contract guarantees that enough information is stably recorded so that it can continue execution after a system crash, with agreement as to whether the interaction occurred and the message that was exchanged.

Each element of a distributed system is characterized with a component type.

1. **Persistent Component (Pcom):** a component whose state is guaranteed to survive system crashes, the execution result being as if the crash did not occur;
2. **External Component (Xcom):** a component that is outside of the control of our infrastructure and for which we cannot make a guarantee;
3. **Transactional Component (Tcom):** a component that executes transactions, and that guarantees that the effects of committed transactions will persist and the effects of uncommitted transactions will be erased.

Between persistent components and each component type, there is a particular flavor of interaction contract that needs to be honored for persistence and exactly once execution to be assured. *But first—an admission.* No system can guarantee failure masking under all circumstances. For example, in a system interaction with a human user, if the system crashes before all input from the user is stored stably (usually by logging), then the user will need to re-enter the input. Despite this, the "interaction" can be "exactly once". And we can minimize the window of vulnerability, and that is what an external interaction contract does. For the other interactions, we "shut" the window.

As there are three component types, we also have three flavors of interaction contracts.

1. **Committed Interaction Contract (CIC):** between persistent components;
2. **External Interaction Contract (XIC):** between external component and persistent component;
3. **Transaction Interaction Contract (TIC):** between persistent component and transactional component.

2.2 Committed Interaction Contract

I'll describe a committed interaction contract here and refer you to [5] for external interaction contracts. Transaction interaction contracts, which underlie our web services contract, are described in the next section.

A **committed interaction contract** between two Pcoms consists of the following obligations:

Sender Obligations:

- **S1: Persistent State**: The sender promises that its state at the time of the message or later is persistent.
- **S2: Unique Persistent Message**
  - **S2a:** Sender promises that each message is unique and that it will send message repeatedly until receiver releases it from this obligation (R2a)
  - **S2b:** Sender promises to resend the message upon explicit receiver request until receiver releases it from this obligation (R2b).

Sender obligations ensure that an interaction is **recoverable,** i.e. it is guaranteed to occur, though not with the receiver guaranteed to be in exactly the same state.

Receiver Obligations:

- **R1: Duplicate Message Elimination:** Receiver promises to eliminate duplicate messages (which sender may send to satisfy S2a).
- **R2: Persistent State**
  - **R2a:** Receiver promises that before releasing sender obligation S2a, its state at the time of message receive or later is persistent. The interaction is now **stable**, i.e. the receiver knows that the message was sent.
  - **R2b:** Receiver promises that before releasing the sender from obligation S2b, its state at the time of the message receive or later is persistent without the need to request the message from the sender. The interaction is now **installed,** i.e., the receiver knows that the message was sent, and its contents.

CIC's ensure that the states of Pcoms not only persist but that all components "agree" on which messages have been sent and what the progress is in a multi-component distributed system. S2 and R1 are essentially the reliable messaging protocol for assuring exactly once message delivery, coupled with message persistence, while S1 and R2 provide persistent state for sender and receiver. Reliable messaging is not sufficient by itself to provide robust applications. Also, the CIC describes the requirements more abstractly than a protocol in order to maximize optimization opportunities. For example, it is not necessary to log an output message to persist it if the message can be recreated by replaying the application from an earlier state.

In our Phoenix/App prototype [2, 3], our infrastructure intercepts messages between Pcoms. It adds information to the messages to ensure uniqueness, resends messages, and maintains tables to eliminate duplicates; it logs messages as needed to ensure message durability. Message durability permits Phoenix to replay components from the log, so as to guarantee state persistence (see, e.g. [3]). Because the interception is transparent, the application executing as a Pcom needs no special provisions in order to be persistent, though some limitations exist on what Pcoms can do.

2.3 Transaction Interaction Contract

A transaction interaction contract[1] is a model for how we deal with web services. Thus we describe it in detail. The TIC explains how a Pcom reliably interacts with, e.g., transactional database systems- and what a transactional resource manager (e.g. DBMS) needs to do, not all of which is normally provided, so as to ensure exactly once execution.

When a Pcom interacts with a Tcom, the interactions are within the context of a transaction. These interactions are of two forms:

- **Explicit** transactions with a "begin transaction" message followed by a number of interactions within the transaction. A final message (for a successful transaction) is the "commit request", where the Pcom asks the Tcom to commit the work that it has done on behalf of the Pcom.
- **Implicit** transactions, where a single message begins the transaction and simultaneously requests the commit of the work. This message functions as the "commit request" message.

For explicit transactions, the messages preceding the commit request message do not require any guarantees. Should either Pcom or Tcom crash, the other party knows how to "forget" the associated work. The components do not get confused between a transaction that started before a crash, and another that may start after the crash. The later is a new transaction. Further, at any point before the "commit request" message, either party can abandon the work and declare the transaction aborted. And it can do so unilaterally and without notice. The only Tcom obligation is the ability to truthfully reply to subsequent Pcom messages with an "I don't know what you are talking about" message (though more informative messages are not precluded).

At the end of a transaction that we want to commit, we re-enter the world where contracts are required. Interactions between Pcom and Tcom follow the request/reply paradigm. For this reason, we can express a TIC in terms that include both the request message and the reply message. We now describe the obligations of Pcom and Tcom for a transaction interaction contract initiated by the Pcom's "commit request" message.

Persistent Component (Pcom) Obligations:

PS1: Persistent Reply-Expected State. The Pcom's state as of the *time at which the reply to the commit request is expected*, or later, must persist without having to contact the Tcom to repeat its earlier sent messages.

- The persistent state guarantee thus includes the installation of all earlier Tcom replies within the same transaction, e.g., SQL results, return codes.
- Persistence by the Pcom of its reply-expected state means that the Tcom, rather than repeatedly sending its reply (under TS1), need send it only once. The Pcom explicitly requests the reply message, should it not receive it, by resending its commit request message.

---

[1] The form of transaction interaction contract presented here is the more complete specification given in the TOIT paper [4].

PS2: Unique Persistent Commit Request Message: The Pcom's commit request message must persist and be resent, driven by timeouts, until the Pcom receives the Tcom's reply message.

PR1: Duplicate Message Elimination: The Pcom promises to eliminate duplicate reply messages to its commit request message (which the Tcom may send as a result of Tcom receiving multiple duplicate commit request messages because of PS2).

PR2: Persistent Reply Installed State: The Pcom promises that, before releasing Tcom from its obligation under TS1, its state at the time of the Tcom commit reply message receive or later is persistent without the need to request the reply message again from the Tcom.

Transactional Component (Tcom) Obligations:

TR1: Duplicate Elimination: Tcom promises to eliminate duplicate commit request messages (which Pcom may send to satisfy PS2). It treats duplicate copies of the message as requests to resend the reply message.

TR2: Atomic, Isolated, and Persistent State Transition: The Tcom promises that before releasing Pcom from its obligations under PS2 by sending a reply message, that it has proceeded to one of two possible states, either committing or aborting the transaction (or not executing it at all, equivalent to aborting), and that the resulting state is persistent.

TS1: Unique Persistent (Faithful) Reply Message: Once the transaction terminates, the Tcom replies acknowledging the commit request, and guarantees persistence of this reply until released from this guarantee by the Pcom. The Tcom promises to re-send the message upon explicit Pcom request, as indicated in TR1 above. The Tcom reply message identifies the transaction named in the commit request message and faithfully reports whether it has committed or aborted.

A TIC has the guarantees associated with reliable message delivery both for the commit request message (PS2 and TR1) and the reply message (PR2 and TS1). As with the CIC, these guarantees also include message persistent. In addition, in the case of a commit, both Pcom (PS1) and Tcom (TR2) guarantee state persistence as well. As with CIC, we stated TIC requirements abstractly.

2.4 Comparison with Transaction Processing

Unlike persistent components, traditional transaction processing applications are stateless. That is, there is no meaningful state outside of a transaction except what is stored explicitly in a queue or database. And, in particular, there is no active execution state. Each application "step" is a transaction; the step usually is associated with the processing of a single message.

A typical step involves reading an initial state from a queue, doing some processing, updating a database, digesting a message, and writing a queue (perhaps a different queue), and committing a distributed transaction involving queues, databases, and message participants. This requires 2PC unless read/write queues are supported by the same database that is also involved. When dealing with distributed TP, queues are frequently different resource managers, and so is the database. So typically, there are at least two log forces per participant in this distributed transaction.

Because all processing is done within a transaction, handling transaction failures requires special case mechanisms. Typically, a TP monitor will retry a transaction some fixed number of times, hoping that it will succeed. If it fails repeatedly, then an error is posted to an administrative (error) queue to be examined manually. The problem here is that no application logic executes outside of a transaction. Program logic fails when a transaction fails. So how does an application understand what happened to its request if a reply is enqueued on an error queue?

There is, not unexpectedly, a relationship between interaction contracts and distributed transactions. Both typically require that logs be forced from time to time so as to make information stable. But the interaction contract is, in fact, the more primitive notion. And, if one's goal is application program persistence, only the more primitive notion is required. Only if rollback is needed is the full mechanism of a transaction required. Further, in web services, as for workflow in general, "rollback" (compensation) is frequently separate from the transaction doing the forward work. For distributed, web services based computing, we do not believe that the tight coupling and high overhead needed for transactions will make them the preferred approach.

There is an even deeper connection between interaction contracts and 2PC. The message protocols in two phase commit are, in fact, instances of transaction interaction contracts. However, by unlocking the TIC from the commit coordination protocol, we make it possible to have a better, more flexible, efficient, and opaque end-to-end protocol for making applications robust. It is these properties that make interaction contracts well suited for web services applications. We'll see this in the next section.

3   Opaque Web Services Using Interaction Contracts

3.1   Web Services Characteristics

The setting for web services is quite different from traditional transaction processing. The TP world was usually completely within an enterprise, or where that wasn't the case, between limited clients and a service or services within a single enterprise. But web services are intended specifically for the multi-enterprise or at least multi-organizational situation. Site autonomy is paramount. This is why web services are "arms length" and opaque, based on messages, not RPC. This is why the messages are self-describing and based on XML.

Because web services are opaque, a transaction interaction contract is not quite appropriate. An application is not entitled to know whether a web service is performing a "real" transaction. It is entitled to know only something about the end state. Further, a web service provider will be very reluctant to enter into a commit protocol with applications that are outside of its control. But web services need to be concerned about robust applications. In particular, a web service should be concerned with enabling exactly-once execution for applications. This is the intent of the web services interaction contract.

3.2 Interacting with a Web Service

There can be many application interactions with a web service that require no special guarantees. For example, an application asks about the price and availability of some product. Going even further, a customer may be shopping, and have placed a number of products in his shopping cart. There is, at this point, no "guarantee" either that the customer will purchase these products, or that the web service will have them in stock, or at the same price, when a decision is eventually made. While remembering a shopping cart by the web service is a desirable characteristic, it is sufficient that this remembering be a "best effort", not a guarantee. Such a "best effort" can be expected to succeed well over 99% of the time. Because of this, it is not necessary to take strenuous measures, only ordinary measures, to do the remembering. But a guarantee must always succeed, and this can require strenuous efforts. We want, however, to reserve these efforts for the cases that actually require them.

By analogy with the transaction interaction contract, guarantees are needed only when work is to be "committed". As Tcom's can forget transactions before commit, so web services can forget units of work. An application can likewise forget this unit of work. No one has agreed to anything yet. There is no direct way an application can tell whether a work unit is a transaction, or not. (Of course, it might start other sessions with other work units and see whether various actions are prevented, and perhaps infer what is going on.) Whether there is a transaction going on during this time is not part of any contract.

A web service may require that an application remember something, e.g. the id for the unit of work, as it might for a transaction, or as it might for a shopping session with a particular shopping cart. But this is not a subject of the guarantee. The guarantee applies exactly to the message in which work is going to be "committed". Everything up to this point has been "hypothetical". Further, if the web service finds the "commit request" message not to its liking, it can "abort". Indeed, it can have amnesia in any circumstance, and when that happens, there is no guarantee that the web service will remember any of the prior activity.

3.3 A Web Services Interaction Contract

It is when we get to the "final" message, e.g. when a user is to purchase an airline ticket, that we need a guarantee. In a web services interaction contract (**WSIC**), only the web service makes guarantees. If the application also takes actions of the sort required of a Pcom in the transaction interaction contract, then it can ensure its persistence. Without the WSIC, it would not be possible to implement Pcoms interacting with web services. But the web service contract is independent of such an arrangement. Thus a WSIC is a *unilateral* pledge to the outside world of its applications. Nothing is required of the applications.

The WSIC requirements for the web service resemble the requirements imposed on transactional components by a TIC.

Web Service (WS) Obligations:

WS1: Duplicate Elimination: WS promises to eliminate duplicate commit request messages. It treats duplicate copies of the message as requests to resend the reply message.

WS2: Persistent State Transition: The WS promises that it has proceeded to one of two possible states, either "committing" or "aborting" and that the resulting state is persistent. [These states are in quotation marks because there may not be a connection with any particular transaction.]

WS3: Unique Persistent (Faithful) Reply Message: WS awaits prompting from the application to resend, accomplished by the application repeating its commit request message. Once the requested action terminates, the WS replies acknowledging the commit request, and guarantees persistence of this reply until released from this guarantee. The WS promises to resend the message upon explicit request. Message uniqueness permits an application to detect duplicates.

An important aspect of the WSIC is that it says nothing about how the web service meets its WSIC obligations. This is unlike the TIC, where the transaction component is required to have an atomic and isolated action (transaction). Thus, a web service might meet its requirements using perhaps a persistent application, or a workflow. It might exploit transactional queues or databases or file systems. This is the other side of the value of the opaque interaction contract. It does not prescribe how a web service meets its obligations, it only describes the obligations. This obligation ensures that the request to the web service is "idempotent". Note that the WS action taken, like a logged database operation, does not need to be idempotent, and in general is not. It is the web service that provides idempotence, i.e. exactly once execution, by ensuring that the action is only executed once. We discuss next how an application can use a WSIC to ensure robust behavior.

3.4 Robust Applications

What the WSIC guarantees, as with the TIC, is that **if** an application is written as a persistent component, following the rules for the Pcom in the TIC, then the application can be made to survive system crashes. We showed how Pcoms can be made persistent previously [2, 3, and 4].

The key role that the WSIC plays is to ensure that an interaction can be re-requested should a failure occur during a web service execution. In this case, the WSIC ensures that the web service activity is executed exactly once, despite potentially receiving multiple duplicate requests (WS1), that the web service "committed" state, once reached is persistent (WS2), and that the reply message will not be lost because it is persistent, and it is unique to ensure that it cannot be mistaken for any other message (WS3), so that duplicates can be eliminated by the application.

We have argued before that programming using persistent applications is easier and more natural than programming using stateless applications. A stateful persistent application need not be arranged into a "string of beads" style, where each "bead" is a transaction that moves the state from one persistent queue to another. Rather, the application is simply written as the application logic demands, with persistence provided by logging and by the ability to replay crash interrupted executions [3]. And it is this program logic that can deal with errors, either exercising other execution paths or at least reporting errors to end users, or both.

4 An Example

4.1 The Application

In this section, we explore ways to implement a web service that satisfy the WSIC. This will illustrate how the flexibility permitted by an opaque web service can be exploited for both implementation ease and to enable persistent applications.

Our application is a generic order entry system. We do not describe it in detail. But it responds to requests about the stock of items it sells, it may permit a user to accumulate potential purchases in a "shopping cart". And, finally, when the client (application or end user) decides to make a purchase, the "commit request" for this purchase is supported by a web service interaction contract that will guarantee exactly once execution.

4.2 Using Transactional Message Queues

A "conventional" transaction processing method of implementing our web service might be to use message queues [1, 7]. When an application makes a request, the web service enqueues the request on its message queue. This executes as a transaction to ensure the capture of the request. The application must provide a request id used to uniquely identify the work item on the queue. The message queue permits only a single instance of a request or reply with a given request id.

The request is executed by being dequeued from the message queue, and the order is entered into the order database, perhaps inventory is checked, etc. When this is complete, a reply message is enqueued to the message queue. This reply message may indicate the order status, what the ship date might be, the shipping cost and taxes, etc. This is the WS2 obligation and part of the WS3 obligation, since both the state and the message are guaranteed to persist.

The application resubmits its request should a reply not arrive in the expected time. The web service checks the message queue for the presence of the request. If not found, the request is enqueued. If it is found, the web service waits for the request to be processed and the reply entry available. If the reply is already present, it is returned directly to the application. The queue has made the reply durable. The web service ensures that request and reply are unique so that duplicates are eliminated, satisfying WS1.

Note here that if the application executes within a transaction, as is the case for traditional transaction processing, and its transaction fails, there is no convenient place to handle the failure. But here, even when the web service uses conventional queued transaction processing, the application can be a persistent one. When that is the case, application logic can handle web service failures. The application is indifferent to how the web service provides the WSIC guarantees, only that they are provided, enabling application persistent.

4.3 Another Approach

We can change the implementation of the web service to provide the WSIC guarantees in a simpler and more efficient manner if the web service has the freedom to modify database design. We guarantee duplicate elimination, a persistent state transition and a persistent output message all by adding a request id field to each order in our order database. We enter an order in the order table using a SQL insert statement. We define this table with a uniqueness constraint, permitting only one order with the given request id to be entered.

Should our activity in the web service be interrupted by a system crash, then there are a number of possible cases.

1. We have no trace of the request because the transaction updating the order table was not committed prior to the crash. In this case, it is as if we have not seen the request. A "persistent" application will resend the request.
2. We have committed the transaction that updates the order table. Subsequent duplicate requests are detected when we again try to insert an order with the given request id into the order table. The duplicate is detected, the transaction is bypassed, and the original reply message is generated again based on the order information in the table.

The implementation strategy we sketch here avoids the need to have the persistent message part of the WSIC released explicitly. The idea is that the request id remains with the order for the entire time that the order is relevant. With the traditional transaction processing approach, the release for the persistence guarantee is done with the commit of a dequeuing operation for the reply on the message queue.

The bottom line here is that the WSIC provides abstract requirements. The web service can decide how to realize them. Message queues are one way of doing this. But, as can be seen, there can be other, perhaps more effective, approaches.

4.4 The Web Service Client

Because the WSIC is an abstract, opaque characterization of requirements for a web service, an application program using the web service can essentially do whatever it wants, since the WSIC is a unilateral guarantee by the web service. The application has no obligations under the WSIC.

If the application doesn't do anything special, the program state will not be persistent across system crashes. But the WSIC guarantees are useful in any case. With respect to our order entry system example, for instance, it is surely useful for an end user to be able to ask about an order's status. And having a persistent state reflecting the order is usually considered a minimal requirement for business data processing needs.

If the application wants to realize exactly once semantics for its request, then the WSIC enables this robustness property to be realized. Not surprisingly, the application needs to implement the Pcom side of the transaction interaction contract (TIC). That is, the application becomes persistent when it assumes the obligations of the persistent component in a TIC.

5 Discussion

A number of additional subjects are worth mentioning briefly.

5.1 Undo Actions for Business Processes

To construct long duration workflows, it is usually a requirement that some form of compensation action be possible for each forward action of the workflow. To support this kind of scenario, we can associate with each "action" of a web service a "cancel" (or undo) action. This says nothing about atomicity. The "cancel" activity can be as opaque as the original action. It says nothing about the details of the "inverse" action. But it puts the responsibility for the cancel (or undo) action on the web service, which, after all is the only autonomous entity really capable of doing it.

One way of dealing with this is to submit the same id used for the commit request with a cancel request (which itself is a form of "commit request" obeying the TIC obligations). There may be a charge assessed for this under some circumstances, and that should be part of the WSDL description of the web service. The response message to this cancel request should be something like "Request cancelled". This is independent of whether the "forward" request was ever received or executed since once the request is cancelled, there should be no requirement that the web service remember the original request.

By supporting a "cancel request", a web server enables an application to program a compensation action should the application need to "change plans". Note that this says nothing about how the application figures out what needs undoing, etc. Again, web services are opaque. But by providing a "cancel request", they enable an application to be written that undoes earlier work as appropriate.

There need be no requirement that a web service provide a "cancel request". But there is no requirement also, that an application program use any specific web service. But many e-commerce sites support canceling orders, and many web services should be willing to support "cancel request", especially if it were possible to charge for it.

A "cancel" request may only be a best efforts cancellation, e.g. the canceling of an order to buy or sell shares of stock. If the cancellation fails, then there is an obligation to faithfully report that failure. In this case, the web service is obligated to maintain the original action state so as to be able to generate a persistent "cancellation failed" message.

5.2 Releasing Contracts

Persistent states and persistent message obligations of the parties to an interaction contract may require an eventual release. Application programs that exploit the obligations eventually terminate. An order is eventually filled and at that point becomes of historical interest but not of current interest. We have not discussed up to this point how the contracts might be released. For web services, recall, the obligations are unilateral and apply only to the web service. Any application effort to exploit the WSIC, e.g. to provide for its persistence, is purely at the its own discretion. Below we discuss some alternatives for the web service.

No Release Required: This is both very useful and very simple for applications. It means that no matter how long the application runs, the web service will retain the information needed to effectively replay the interaction. A variant of this is that, for example, once an order is shipped, the shipped object ends the WSIC obligations. So it is frequently possible to remove information about old orders (old interactions) from the online system without compromising the WSIC guarantees, and without requiring anything from the application.

Release Encouraged: In this scenario, when an application releases the web service from its WSIC obligations, the web service can remove information associated with the interactions from the online system. If only a small number of apps do not cooperate, that will not be a major issue. Storage is cheap and plentiful. A further step here is to "strongly" encourage release, e.g. by giving cooperators a small discount. Another possibility is to eventually deny future service until some old WSIC's are released. This distinguishes well-behaved applications from rogues, and eventually limits what the rogues can do.

Release Required: When web service providers feel that it is too much of a burden on them to maintain interaction information in their online system, an application can be required to release the contract. A frequent strategy is to stipulate that the contract is released at the time of the next contact or the next commit request. If there is no such additional contact that flows from the application logic, then release can be done at the time the application terminates (or its session terminates). Of course, stipulations can be ignored. So a further "clause" in the web service persistence guarantee might "publish" a time limit on the guarantee. For example, one might safely conclude that most applications requesting a web service would be complete within four hours of the commit request having been received, or within one day, etc. This permits "garbage collecting" the information that is older than the published guarantee.

5.3 Optimizations

The existence of web services supporting WSIC's can frequently enable persistence for applications more efficiently than can traditional transaction processing. Typically, with traditional transaction processing, each message exchange is within a separate transaction. Input state on a queue and one message are consumed, and the output state is placed on another queue, then the transaction is committed. Thus there are two log forces per resource manager (message source, input queue, and output queue) and at least one, perhaps two, for the transaction coordinator.

Using interaction contracts, we have an opportunity to avoid multiple forced log writes. The application may have several interactions, each with a different web service. Assuming that contract release is either not required, or occurs on the next interaction, it is not necessary for the application to log each of these interactions immediately to make them stable, as is done by committing transactions in classic TP. These interactions are stable via replay, using the sender's stable message. This results in many fewer forced log writes. Each web service may require a log force, and eventually the application will need to force the log, but this can be amortized over multiple web services. So, "asymptotically, we might have as few as one log force per method call instead of several.

5.4 Summary

We have shown how robust web based applications can be enabled by web services that meet the relatively modest requirements for a unilateral web services interaction contract. The WSIC is an opaque requirement in that it does not specify (or even reveal) how it is that the web service satisfies the WSIC. Further, it usually permits robust, i.e. persistent applications, to be realized at lower cost than traditional transaction processing. Finally, the application can be a stateful one. This has two advantages: (i) it is a more natural programming style than the "string of beads" style required by traditional TP; (ii) it enables program logic in the persistent application to deal with transaction failures, something that is not easily accommodated in traditional TP.

References

1. P. Bernstein, and E. Newcomer. *Principles of Transaction Processing*. Morgan Kaufmann, 1997.
2. R. Barga, S. Chen, D. Lomet. Improving Logging and Recovery Performance in Phoenix/App, *ICDE* (March 2004) 486–497.
3. R. Barga, D. Lomet, S. Paparizos, H. Yu, and S. Chandrasekaran. Persistent Applications via Automatic Recovery. *IDEAS* (July, 2003) 258–267.
4. R. Barga, D. Lomet, G. Shegalov, and G. Weikum. Recovery Guarantees for Internet Applications *ACM TOIT* 4(3) (August, 2004) 289–328.
5. R. Barga, D. Lomet, and G. Weikum. Recovery Guarantees for General Multi-Tier Applications. *ICDE* (March 2002) 543–554.
6. E.N. Elnozahy, L. Alvisi, Y. Wang, and D.B. Johnson. A Survey of Rollback-Recovery Protocols in Message-Passing Systems. *ACM Comp. Surv.* 34(3), 2002.
7. J. Gray and A. Reuter. *Transaction Processing: Concepts and Techniques*. Morgan Kaufmann, 1993.
8. Microsoft. Microsoft .Net Framework Developer Center.   http://msdn.microsoft.com /netframework/
9. Oasis. Universal Description, Discovery and Integration of Web Services http://www.uddi.org/specification.html
10. D. Parnas. On the Criteria To Be Used in Decomposing Systems into Modules. *CACM* 15(12): (December 1972) 1053–1058.
11. WC3. Web Services Description Language (WSDL) 1.1 http://www.w3.org/TR/wsdl .
12. WC3. Web Services Coordination. 106.ibm.com/developerworks/library/ws-coor/.

When are Two Web Services Compatible?

Lucas Bordeaux, Gwen Salaün, Daniela Berardi, and Massimo Mecella

DIS, Università di Roma *"La Sapienza"*, Italy **
`lastname@dis.uniroma1.it`

Abstract. Whether two web services are compatible depends not only on static properties like the correct typing of their message parameters, but also on their *dynamic* behaviour. Providing a simple description of the service behaviour based on process-algebraic or automata-based formalisms can help detecting many subtle incompatibilities in their interaction. Moreover, this compatibility checking can to a large extent be automated if we define the notion of compatibility in a sufficiently formal way. Based on a simple behavioural representation, we survey, propose and compare a number of formal definitions of the compatibility notion, and we illustrate them on simple examples.

1   Introduction

Performing complex tasks typically requires to make a number of web services work together. It is therefore necessary to ensure that these services will be able to interact properly, which calls for a clear understanding of the notion of *compatibility*. Compatibility is moreover closely related to another problem, *substitutability*: when can one service be replaced by another without introducing some flaws into the whole system? Ensuring the substitutability of a previously-used service by a new one is necessary in many situations, for instance, when the old service comes to be temporarily unreachable, or when a new release of a service is proposed which provides better functionalities, better Quality of Service, or has lower cost [AMPP03]. This paper is an attempt to address the issue of compatibility and to show how this notion can also be used to tackle substitutability issues.

Incompatibilities between WSs can arise at a number of different levels. For instance two services can be incompatible because the messages they can send and receive (as declared, say, in their WSDL interface) have incompatible types. Such a *static* compatibility is essential to check, but a more challenging problem is raised by the very *dynamic* nature of the WS interaction, which is based on an exchange of messages which can be ordered in complex sequences. We would like

---

** L. Bordeaux and G. Salaün are partially supported by project ASTRO funded by the Italian Ministry for Research under the FIRB framework (funds for basic research). D. Berardi and M. Mecella are partially supported by MIUR through the FIRB project *MAIS* (`http://www.mais-project.it`).

M.-C. Shan et al. (Eds.): TES 2004, LNCS 3324, pp. 15–28, 2005.

to ensure that, whatever scenario the interaction goes through, undesired situations (messages not received, impossibility to terminate the interaction, *etc.*) will never occur. This requires reasoning on the *behavioural* features of WSs, *i.e.*, to describe and examine the possible sequences of messages each of them can send or receive. Note that other reasons for incompatibility can exist, for instance *semantic* incompatibilities (a car-renting service can hardly be considered compatible with a client which wants to buy music online). The paper shall focus only on behavioural aspects, assuming that the names of the exchanged messages are standardised and that semantic compatibility is guaranteed.

The need for WSs to provide a publicly-available interface describing their behaviour has been recognised by the community, especially for choreography issues. To give precise definitions of the notions of compatibility and substitutability, and to design and implement automated tools for checking these properties, it is convenient to abstract away from the XML syntax and to choose a simple formalism to represent behaviour. Whereas some authors have used Petri nets [NM02] or process algebra [MB03, SBS04] for instance, we use here Labelled Transition Systems, which are in a sense a simple model underlying all these richer notations, and which also faithfully correspond to the constructs found in proposals for behavioural interfaces like WSCI. This framework enables us to give a uniform presentation of different notions of compatibility of behavioural interfaces, inspired in particular from the area of software components [YS97, BPR02, BBC02, CFP+03], to discuss them and to illustrate them on simplistic yet easily generalisable examples. For simplicity, we focus on definitions where only 2 services interact and we discuss the generalisation to an unbounded number of services in the end of the paper. Recent related work in the WS literature [BCPV04, MPC01, Mar03, LJ03] was typically informal or focused on very specific representations, or related to non-behavioural features (see, e.g., the work based on ontology [AMPP03]). Moreover, a number of different notions emerged in this literature, and our present work is, to the best of our knowledge, the first attempt to classify them and address the problem in a systematic way.

The organisation of this paper is as follows. Section 2 introduces the formal model of WSs we come up with to formalise afterwards the compatibility notion. Section 3 tackles the notion of compatibility while Section 4 deals with substitutability. We draw up in Section 5 some concluding remarks and perspectives.

2   Our Model for Web Services

The behaviour of a web service will be represented as a *Labelled Transition System* (LTS) as illustrated by the (simplistic) service shown in Fig. 1, which expects to receive a request, and can either send an error message if the requested product is not available, or send more info on the shipping conditions and wait for a confirmation/cancellation before terminating the interaction.

More generally, services will be represented by a set S of *states* (the circles), *transitions* between the states (the arrows) and *actions* (the labels of the arrows)

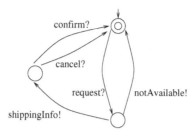

Fig. 1. A Labelled Transition System modelling a simple service

which can either be *emissions* or *receptions*. Sending a message n is written $n!$ and receiving it is written $n?$, and we denote by A the set of actions of the form $n!$ or $n?$ (where n ranges over a predefined set of message names). The transitions are specified by a (partial) function t which, given a state $s \in S$ and an action $a \in A$ specifies to which new state $s' \in S$ the service moves (we note $t(s, a) = \bot$ when there is no transition labelled by action a going out of state s). To complete the picture we need to identify the initial state $i \in S$ and the set of final states $F \subseteq S$ (double circles) in which the job that the service has to realize is considered finished. Since we deal with two services, we use subscripts like S_1, t_1, i_1 or F_1 (*resp.* S_2, *etc.*) to make explicit that we consider the states, the transition function, the initial or final states, of service 1 (*resp.* service 2). Note that our model is close to other proposals of the literature [BCG$^+$03, BFHS03] in that:

- the receiver of messages is implicit because we deal with 2 services, messages are necessarily sent to the partner;
- for the sake of simplicity, we do not represent the data carried by the messages because we focus on purely behavioural features;
- it is deterministic: one action applied in a given state leads to a completely determined state.

Note that the terminology is somewhat subtle here: we do have some non-determinism in that we reflect *all* possible executions of the represented system, and we therefore make implicit use what the process algebra community would call *non-deterministic choice* [Mil89]. We are deterministic, though, in the automata-theoretic sense, which is that no two actions labelled by the same name can be applied in one given state (*i.e.*, the transition relation is indeed a *function*). It makes little sense in practice, starting from some particular state s, to say that a particular action (say, the reception of message "confirm") will lead our service *either to state s_1 or to state s_2*. Such a modelling could be "determinised" using classical methods from automata theory. If needed however, our model and definitions could in any case easily be generalised to the non-deterministic case.

Determining compatibility and substitutability of services requires to reason on the possible scenarios which their interaction can go through. We therefore define formal material to model the sequence of messages which can be sent by one service and received by the other. These definitions correspond to a *synchronous* two-party communication model. It makes sense in some contexts to use asynchronous communications instead (at the price of much more severe difficulties in the definitions and verification algorithms), but the synchronous communication model is normally exploited in current WS technology, see for instance choreography description languages. The differences which arise when we consider an asynchronous model are briefly illustrated in section 5.

2.1   Reasoning on Possible Scenarios

A (finite) sequence of actions is written using a list notation $[m_1; m_2; \ldots]$, where each $m_i \in A$. In particular, $[]$ represents the empty list. We use the notation $s \xrightarrow{l} s'$ to express the fact that, starting from state s, the considered service can perform the sequence of actions specified by the list l, which leads it to state s'.[1]

When a service sends (*resp.* receives) message m, this means that the other service simultaneously evolves by receiving it (*resp.* sending it). We use the notation $\overline{m}$ to represent the *opposite* action of action m, i.e., we define $\overline{n?}$ as $n!$ and $\overline{n!}$ as $n?$. We generalise this notation to sequences, defining the list $\overline{[m_1; m_2; \ldots]}$ as $[\overline{m_1}; \overline{m_2}; \ldots]$. The valid scenarios which may occur can therefore be defined as follows: service 1 (which starts from the initial state i_1) can perform the sequence of actions $l = [m_1; m_2; \ldots]$ and reach state s_1, if and only if service 2 can perform the sequence of actions $\overline{l}$ leading to some valid state s_2. In other words, a sequence l is a *valid sequence of actions* for service 1 iff we have:

$$i_1 \xrightarrow{l} s_1 \quad \text{and} \quad i_2 \xrightarrow{\overline{l}} s_2$$

A pair of states $\langle s_1, s_2 \rangle$ for which there exists such a valid sequence of actions l is called a *reachable pair of states*. Reachable pairs of states are important when reasoning on behavioural issues since they represent the possible configurations the system can reach in every possible scenario of interaction.

One last thing we need is to reason on the set of messages which can be sent and received by a service in a particular state, hence it is useful to define:

- *emissions$_i$(s)* as the set containing the names of the messages which service i can send when it is in state s, i.e., $\{n \mid t_i(s, n!) \neq \bot\}$;
- *receptions$_i$(s)* as the set containing the names of the messages which service i can receive when it is in state s, i.e., $\{n \mid t_i(s, n?) \neq \bot\}$.

---

[1] Formally, predicate $s \xrightarrow{l} s'$ is defined by:

$$s \xrightarrow{[]} s \quad \text{and} \quad s \xrightarrow{[m_1; m_2; \ldots]} s'' \ \text{ iff } \ t(s, m_1) = s' \text{ and } s' \xrightarrow{[m_2; \ldots]} s''.$$

3   Notions of Compatibility

From a behavioural viewpoint, intuition suggests that *"two services are compatible if they can interact properly"*. In the following we describe 3 possible definitions of this informal notion.

3.1   Opposite Behaviours

The first notion of compatibility, and perhaps the most natural to start with, is that of *opposite behaviours*. It is based on the observation that, when a service emits something, the other should receive it, so in a sense *the behaviour of service 2 should be the same as service 1, but with receptions instead of emissions, and vice-versa.*

Defining the *opposite behaviour* $\overline{A}$ of a service A as the service obtained when the emissions are changed to receptions and vice-versa (*i.e.*, the transition function $\overline{t}$ of $\overline{A}$ is defined by: $\overline{t}(s, m!) = t(s, m?)$ and $\overline{t}(s, m?) = t(s, m!)$), we have the following notion:

Definition of Compatibility 1. *Two services A and B are compatible if they have* opposite behaviours, *i.e., A is equivalent to $\overline{B}$.*

The relation is symmetric: A is equivalent to $\overline{B}$ if B is equivalent to $\overline{A}$ for any decent relation of equivalence[2]. But what do we mean by two services being *equivalent*? There are a number of notions of equivalence between processes or labelled transition systems but, since we are in a simple, deterministic case, the most important of these notions (like bisimulation and trace equivalence [Mil89]) are actually equivalent to the following, simplified definition:

Definition 1. *[Observational indistinguishability] Service 1 in state s_1 is observationally indistinguishable from service 2 in state s_2 if:*

- *either the two states are initial (resp. final) or none of them is, and*
- *the same messages can be sent and received in both states and if they lead to states which are observationally indistinguishable:*
  - $t_1(s_1, m) = t_2(s_2, m)$, *and*
  - *if $t_1(s_1, m) \neq \perp$ then $t_1(s_1, m)$ is indistinguishable from $t_2(s_2, m)$.*

The two services are indistinguishable if they are indistinguishable in their initial states.

Just to show that this concept is slightly more subtle than having "identical drawings", the example above shows two indistinguishable services. When interacting with either service starting from its initial state, it is clearly not possible for an external observer to distinguish between the two of them.

---

[2] If $A \equiv \overline{B}$ then $\overline{A} \equiv \overline{\overline{B}} = B$, the only assumptions are that relation $\equiv$ be symmetric and compatible with negation, *i.e.*, that $A \equiv B$ implies $\overline{A} \equiv \overline{B}$.

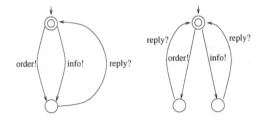

Using this definition of process equivalence, it is interesting to have a look at the set of messages the two services can send and emit. If, in some reachable pair of states, the set of messages one service can emit does not exactly match the set of messages the other can receive, then they can easily be shown to be incompatible. This necessary condition for compatibility also turns out to be sufficient, and we therefore have the alternative way of characterising our first compatibility notion:

Remark 1. Two services are compatible in the sense of Def. 1 if, for any reachable pair of states $\langle s_1, s_2 \rangle$, we have:

$$emissions_1(s_1) = receptions_2(s_2)$$
$$\text{and } emissions_2(s_2) = receptions_1(s_1)$$

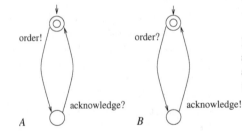

Fig. 2. Two services compatible in the sense of Def. 1: the emissions by one service match exactly the receptions of the other

3.2   Unspecified Receptions

Definition 1 can be seen as too restrictive in some situations [BZ83, YS97]. When two services meet on the web, they may be able to cooperate in a satisfactorily way even when one has slots for receptions which the other one does not intend to use. Consider for instance a service which can receive requests for many types of products and is ready to receive a number of messages including *show_top10_CD_sells*, *search_book_title*, *goto_my_account*, etc. Such services typically provide more than a particular client service really needs, and it makes sense to make them cooperate with a client which can only send a subset of these requests. Compatibility here means that the client will not send requests which the seller cannot satisfy. Of course, the situation is symmetric and it is desired that the client also be ready to accept all the messages sent by the seller—in other words, we want the whole system to have *no unspecified reception*:

Definition of Compatibility 2. *Two web services are compatible if they have no unspecified reception i.e., if, for any reachable pair of states $\langle s_1, s_2 \rangle$, we have that:*

$$emissions_1(s_1) \subseteq receptions_2(s_2)$$
$$and\ \ emissions_2(s_2) \subseteq receptions_1(s_1)$$

Intuitively, the absence of unspecified receptions gives indication that no message will ever be sent whose reception has not been anticipated in the design of the other service. It is immediate to see that this relation is symmetric and that, in particular, a service A is compatible with the service $\overline{A}$ w.r.t. this definition (or to any service indistinguishable from it). This shows that two services compatible in the sense of Def. 1 are also compatible in the sense of Def. 2.

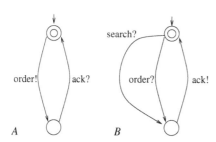

Fig. 3. Two services compatible in the sense of Def. 2 (but *not* compatible according to Def. 1): each service is ready to receive at least all the messages that its mate can choose to send, and possibly more

3.3   Deadlock-Freeness

The two previous definitions have one drawback: they do not consider the question of whether the interaction will reach a final state. In particular, our definitions consider that two services which do not send any message and just do receptions are compatible[3].

Definition 2. *[deadlock] A reachable pair of states $\langle s_1, s_2 \rangle$ is a deadlock if it is impossible from these states to reach a final state.*

The interaction between two services is deadlock-free if no reachable state is a deadlock.

Although a transversal issue, it typically makes sense, when we check the compatibility of two services, to ensure that the resulting application is also deadlock-free. We can therefore use variants of the previous definitions of compatibility where we ensure at the same time the absence of incompatibility (*i.e.,* we guarantee that the service have opposite behaviours or that they have no unspecified reception), and the additional property that terminating the interaction will be possible at any step of the communication.

The notion of deadlock by itself allows to define a third notion of compatibility:

Definition of Compatibility 3. *Two services are compatible if the initial state is not a deadlock, i.e., if there is at least one execution leading to a pair of final states.*

---

[3] This is in a sense perfectly satisfactory: behavioural compatibility is but one among many things one would like to ensure regarding a service-based application; two services may very well be compatible and yet incorrect *w.r.t.* deadlock-freeness or any other correctness criterion.

In other words, we just have to check that one pair of final states belongs to the reachable pairs. This notion guarantees that there is *at least one way* for the two services to interact in a satisfactory way, leading to a final state.

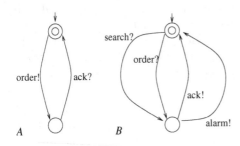

Fig. 4. Two services compatible in the sense of Def. 3 (but not compatible according to Def. 1 & 2): there is at least one possible way of using the two services: the emission of message "order" by the first one will correctly be received by the second, and if the second one acknowledges, the execution can successfully terminate

4   Substitutability

An interesting question closely related to compatibility is *substitutability*: when can we replace a service A by another service A'? In our opinion, the notion can be defined more formally in at least two ways, one application-dependent and one application-independent.

4.1   Context-Dependent Substitutability

The first definition is based on the assumption that we have 2 services and that we remove the first one and replace it by some other service. Substitutability consists of checking whether the substitute is still compatible with the other service which the original service was communicating with.

Definition of Substitutability 1. *In a particular application made of two compatible services A and B, service A' can substitute service A if A' is also compatible with B.*

We therefore have 3 definitions of substitutability, depending on which one amongst the three notions of compatibility we choose. Note that in this definition we have to test the compatibility of the substitute A' with B, *independently* of what the original service A looked like. One natural question is, can we tell by just having a look at A?

The answer is *no* if we use the definitions 2 or 3 of compatibility: there is no way of determining whether "service A' is a substitute of A in the interaction with B" by just having a look at A. Contrary to the first definition of compatibility, the only way is to check the compatibility with B.

On the contrary, the answer is *yes* in the case of the first definition of compatibility: if A and B have opposite behaviours and we want A' to also be compatible with B according to this definition, then A' has to be *observationally indistinguishable* from A:

Remark 2. With the first notion of compatibility, a service can substitute another one iff they are *indistinguishable* (as defined before).

But now it is interesting to note that this definition is *independent of B*. In other words, if we can substitute A by A', the substitute can be used with *any* service B which could interact with A. This gives the idea of the context-independent notion of substitutability developed next.

4.2   Context-Independent Substitutability

As suggested in the last section, we can define a context-independent notion of substitutability as follows:

Definition of Substitutability 2. *A service A' can substitute a service A if it is compatible with* any *service B which is compatible with A.*

This notion of substitutability is helpful, for instance, when we want to replace a service by an updated version which shall be used by many different other services met on the web – *i.e.,* we do not have in advance any information about the services with which we are going to interact. It makes sense to guarantee that the substitute will be able to communicate properly with any service which could work with the original one.

We mentioned in the previous subsection a particular case of this notion of substitutability: if we replace a service A by a service A' which is equivalent (indistinguishable), the substitute will be able to work properly with any service B which could work with A. This defines a notion of context-independent substitutability which corresponds to the first notion of compatibility.

There is no way of using the 3rd notion of compatibility to define a context-independent substitutability: to know that it is possible for service A' to communicate with B in at least one way, we need to fix some particular B. On the contrary, a context-independent substitutability based on definition 2 is possible. To see how, let us consider the following two services:

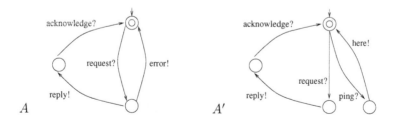

Service A' turns out to be compatible with any service which is compatible with service A according to Def. 2. The reason is that service A' performs *less emissions and more receptions*, so it can properly handle any message that A was designed to receive, and there is no risk of unspecified receptions appearing because of A' sending messages whose reception has not been anticipated in the design of its interlocutor. Intuitively, this corresponds to the case where we have updated service A by adding functionalities: the service can now receive a new type of queries ("ping"), but services which do not use this functionality can still be interact with A'. Service A could also occasionally send an error message which is not needed anymore; the users of service A therefore had a slot for the reception of this message which will not be used, but none of these modifications is a source of incompatibility according to Def. 2. We can define this context-independent substitutability as follows:

Remark 3. With regards to the second notion of compatibility, a service 1 can substitute a service 2 if for any sequence of actions l that is performable in both cases, *i.e.*, such that:

$$i_1 \xrightarrow{l} s_1 \quad \text{and} \quad i_2 \xrightarrow{l} s_2$$

we have:

$$emissions_1(s_1) \subseteq emissions_2(s_2)$$
$$\text{and } receptions_1(s_1) \supseteq receptions_2(s_2)$$

This defines a *partial ordering*. The relation "service 1 is a substitute for service 2" is, in particular, transitive: if we can substitute A by B and B by C, we can also substitute A by C with no problem.

Once again, this definition guarantees that replacing a service by another one will not introduce new unspecified receptions, but does not say anything *w.r.t.* other kinds of flaws. It is interesting for instance to wonder whether this notion of substitutability preserves deadlock-freeness. This is not the case, as shown by the following example:

Consider the example of Fig. 2 (section 3.2) where we had two compatible services A and B. A' can substitute A without introducing unspecified receptions. But the interaction of A' and B cannot reach a final state — we have a deadlock, which was not the case with service A.

The reason for this drawback is that we allow the substitute service to do *less emissions*, while some of these emissions may be needed for the interaction to progress. If we replace the condition in the last definition by:

$$emissions_1(s_1) = emissions_2(s_2)$$
$$\text{and } receptions_1(s_1) = receptions_2(s_2)$$

i.e., if we use a definition of substitutability adapted from our first notion of compatibility, we have a more restrictive notion of substitutability which guarantees two things: 1) the substitute is deadlock-free *w.r.t.* any mate which can work in a deadlock-free manner with the original service; 2) using the substitute does not introduce unspecified receptions.

5   Final Remarks and Future Work

A natural question emerging from our work is whether one notion of compatibility should be preferred. This depends on the problem at hand, and we did not provide a black and white answer. Instead, we have tried to compare the different possibilities and to explain to which situations each applies. In many cases we want two services to work together without unspecified receptions and without deadlock, in which case we can use definition 2 with an additional test for deadlock-freeness to guarantee that the two services will communicate successfully. If a stricter matching of the behaviours is desired, definition 1 should be considered, and if what is needed is simply to see whether an interaction shall be possible, one can use definition 3.

5.1   Variants of the Model

One could consider a number of variants of our model for behaviour representation on which we have based our definitions. In particular, and essentially for the sake of simplicity, we have discussed the questions of compatibility of 2 services, but some definitions can be generalised to situations where an arbitrary number of services interact. This is the case for the notions of deadlock-freeness and of unspecified receptions; it is a bit less clear how the first definition of compatibility can be generalised to more than two services (a service needs to be the opposite of the *composition* of the other services). We also point out the two interesting variants of our model:

Models with Internal and External Non-determinism: Non-determinism in our framework is modelled by the possibility to have several transitions leaving from the same state. We did not distinguish between the two reasons which may lead to such a multiple choice: in some cases, a choice occurs because the service which is considered proposes its partners a number of alternatives among which they can pick freely. This typically occurs when a service is in a state where it is waiting for one among several possible *receptions*; for instance, a book-selling service can in some state wait for either a *buy* message or for *new_search*. On the other hand, choices can also occur because the service itself has to choose among several alternatives. This typically occurs in states where several *emissions* are possible; for instance at some point a bookselling service can either send a *not_available* or a *proceed_order* message. What is modelled here is the "choice" of the bookseller in that it depends on computation it performs internally.

Finer-grained models have been proposed which reflect the distinction between these two sources of non-determinism, which are respectively called *external* and *internal* (see [RR02] for such a work, which discusses a notion of *conformance* related to our *substitutability* and a notion of *stuck-freeness* related to the absence of unspecified receptions). The distinction between the two is important because it makes explicit who has the choice and who should consider all the possible choices of the other. It is important for a book buyer, for instance, to know that the choice between *not_available* and *proceed* is not his.

In most realistic service-based applications though, internal choice appears in our opinion to be synonym of emissions, while external choice is synonym of receptions. We have therefore chosen to base our definitions on the notions of emissions and receptions instead of adding two notations for nondeterminism. Such a notation could be helpful, anyway, if we were to model situations where a number of receptions and emissions are both possible in the same state. Also note that internal non-determinism can alternatively be be represented in some frameworks using *silent actions* (ϵ-transitions, also called "τ actions" in the process algebra literature [Mil89]), which model the fact that a service can at some point silently move to a new state because of internal computations. This provides an alternative way of having both internal and external non-determinism when describing a behaviour.

Models with Asynchronous Communication: Reasoning on *asynchronous communications* raises very subtle problems which we exemplify by the following web services (this example is similar to one exposed in [BZ83]):

If we assume that communication be instantaneous, these two services are perfectly compatible according to every definition we have considered. Now if we consider that messages can be sent asynchronously, a subtle problem can arise: it may happen that service A emits the request and that service B decides to emit the alarm (of which we can think as a timeout sent when service B did not receive any request) before the request arrives to it. The problem is that service A does not expect to receive an alarm signal after having sent its request: this should not occur if we assume synchronous communication, but it might happen if message transmission takes time.

Reasoning on asynchronous communications is unfortunately very often undecidable [BZ83], which means that automated tools are submitted to severe restrictions. Model-checkers like SPIN can verify properties of asynchronous systems in which the transmission delay can be bounded.

5.2   Open Questions and Perspectives

A number of questions are left open and will be considered for future work:

Algorithmic Questions: We did not give the details of the algorithms, but it should be clear that it can be determined automatically whether two services are compatible or if one can substitute another.

Moreover, existing tools developed in the Formal Methods community can be reused. The equivalence between processes can be tested by the bisimulation checkers implemented in many tools based on Process Algebra (the CADP[4] toolbox for the LOTOS language for instance), and the absence of deadlock can be verified using any model-checker. Algorithms to check the absence of unspecified receptions are implemented, for instance, in SPIN[5].

Furthermore, it is important to note that, in the particular framework we have considered, these algorithms provide guarantees in terms of *efficiency*: if we have two services, the runtime of these algorithms is *polynomial* in the size of the two XML or abstract interfaces. We think this is an important prerequisite if we consider services meeting on the web which have to dynamically test their compatibility – most verification algorithms have an at least exponential complexity which severely restricts their wide-scale and fully automated use.

More Sophisticated Notions of Compatibility and Substitutability: In case two services are not compatible, it is natural to try to "do something" to correct the flaws in their interaction. Several authors have investigated the possibility of automatically creating a *patch*, or *adaptator* in order to restore the compatibility. This patch typically consists in an intermediate behaviour which is put between the two services and correct the flaws between their interaction.

This is one of the many possible variants which can be considered to enrich the basic framework we have considered here. Once again, in our opinion, one element which should be taken into account in the computational feasibility of such methods. Typically, the cost of checking these sophisticated notions of compatibility is exponential, which can restrict their applicability to industrial applications.

References

[AMPP03] V. De Antonellis, M. Melchiori, B. Pernici, and P. Plebani. A methodology for e-service substitutability in a virtual district environment. In *Proc. of Conf. on Advanced Information Systems Engineering (CAISE)*, pages 552–567. Springer, 2003.

[BBC02] A. Bracciali, A. Brogi, and C. Canal. Dynamically adapting the behaviour of software components. In *Proc. of Conf. on Coordination Models and Languages (COORDINATION)*, pages 88–95. Springer, 2002.

[BCG+03] D. Berardi, D. Calvanese, G. De Giacomo, M. Lenzerini, and M. Mecella. Automatic composition of e-services that export their behavior. In *Proc. of Int. Conf. on Service-Oriented Computing (ICSOC'03)*, pages 43–58. Springer, 2003.

[BCPV04] A. Brogi, C. Canal, E. Pimentel, and A. Vallecillo. Formalizing web services choreographies. In *Proc. of workshop on Web Services and Formal Methods (WS-FM)*, 2004.

---

[4] http://www.inrialpes.fr/vasy/cadp/
[5] http://spinroot.com

[BFHS03]   T. Bultan, X. Fu, R. Hull, and J. Su. Conversation specification: a new approach to the design and analysis of E-service composition. In *Proc. of World Wide Web Conference (WWW)*, pages 403–410. ACM Press, 2003.

[BPR02]   A. Brogi, E. Pimentel, and A. M. Roldán. Compatibility of Linda-based component interfaces. In *Proc. of workshop on Formal Methods and Component Interaction (FMCI)*, volume 66(4) of *Elec. Notes on Theor. Comput. Science*, 2002.

[BZ83]   D. Brand and P. Zafiropulo. On communicating finite-state machines. *J. of the ACM*, 30(2):323–342, 1983.

[CFP+03]   C. Canal, L. Fuentes, E. Pimentel, J. M. Troya, and A. Vallecillo. Adding roles to CORBA objects. *IEEE Transactions on Software Engineering*, 29(8):242–260, 2003.

[LJ03]   Y. Li and H. V. Jagadish. Compatibility determination in web services. In *Proc. of ICEC eGovernment Services WS*, 2003.

[Mar03]   A. Martens. On compatibility of web services. *Petri Net Newsletter*, 65, 2003.

[MB03]   G. Meredith and S. Bjorg. Contracts and types. *Communications of the ACM*, 46(10):41–47, 2003.

[Mil89]   R. Milner. *Communication and Concurrency*. Prentice Hall, 1989.

[MPC01]   M. Mecella, B. Pernici, and P. Craca. Compatibility of e-services in a cooperative multi-platform environment. In *Proc. of VLDB satellite workshop on Technologies for E-Services (TES)*, pages 44–57. Springer, 2001.

[NM02]   S. Narayanan and S. McIlraith. Simulation, verification and automated composition of web services. In *Proc. of World Wide Web Conference (WWW)*, pages 77–88. ACM Press, 2002.

[RR02]   S. K. Rajamani. and J. Rehof. Conformance checking for models of asynchronous message passing software. In *Proc. of Conf. on Computer Aided Verification (CAV)*, pages 166–179. Springer, 2002.

[SBS04]   G. Salaün, L. Bordeaux, and M. Schaerf. Describing and reasoning on web services using process algebra. In *Proc. of Int. Conf. on Web Services (ICWS)*, pages 43–51. IEEE Computer Society Press, 2004.

[YS97]   D. Yellin and R. Strom. Protocol specifications and component adaptors. *ACM Transactions on Programming Languages and Systems*, 19(2):292–333, 1997.

Negotiation Support for Web Service Selection

Marco Comuzzi and Barbara Pernici

Dipartimento di Elettronica e Informazione-Politecnico di Milano,
Piazza Leonardo da Vinci 32 I-20133 Milano, Italy
comuzzi@elet.polimi.it, barbara.pernici@polimi.it

Abstract. In the literature, the problem of negotiating the characteristics of a web service has been addressed only from the point of view
of the characterization of negotiation protocols, without considering the
problem of coordinating the services that participate in the negotiation
process. We propose a framework for the coordination of different services for web service negotiation during web service selection. We also
discuss the representation of negotiation in process specification and its
enactment.

1 Introduction

Negotiation is the most natural and flexible way to set the QoS parameters
values of a service and its price. Negotiation has been studied since the 50s and
the growth of internet technologies in the last ten years has shifted the interests
of the academic community towards automated negotiation and the study of
best-suited protocols for automating the negotiation process. The first objective
of this field of research was to develop applications with which the user could
interact to participate in online negotiations, major results in this field have been
clinched in the development of online auctions web sites. The next step has been
to develop software agents that can negotiate autonomously. Research in this
field has taken three different perspectives: game theory [18], classical software
agents, and machine learning agents [23]. While game theory is interested in
studying the outcomes of different kinds of negotiation, without focusing on
the protocol itself, many frameworks have been proposed for different kinds of
automated agent-based negotiation, such as auction-based [21], trade-off based
[20] and argumentation based models [22].

The problem of negotiating the characteristics of a service in a dynamic and
loosely coupled environment in a generic Service Oriented Architecture (SOA,
[5]) is becoming an emerging problem in the web service research field. Recent
solutions tried to specify negotiation protocols and messages to be adopted in the
matchmaking process of web services provisioning using ad-hoc XML-based languages, like [12], or consolidated proposals, like BPEL4WS [9], without considering the problem of negotiation as a problem of coordination and orchestration
of different services.

M.-C. Shan et al. (Eds.): TES 2004, LNCS 3324, pp. 29–38, 2005.

We propose a framework for negotiation using a coordinator exploiting the service oriented architecture of the MAIS (Multichannel Adaptive Information Systems)[1] project for the description of service provisioning ([17], [2], [15]).

The paper is structured as follows: in Section 2 we introduce the reference architecture; Section 3 discusses coordination problems involved in service negotiation; Section 4 presents an example of negotiation protocol specification and Section 5 concludes the paper.

2   The MAIS Architecture

The MAIS architecture for service provisioning is reported in Figure 1. Web services are registered in the MAIS registry, an enhanced UDDI registry, which provides functionalities of service discovery specialized by ontology-based descriptions, and are invoked by a Concrete Service Invocator [2].

In the MAIS framework, abstract and concrete services are considered [1]: an abstract service is characterized only by the abstract part of WSDL specification (*type*, *messages*, *operation* and *portType* elements); a concrete service is a binding of the abstract definition to a specific endpoint and a specific communication protocol to invoke the service, for example sending SOAP messages over HTTP. The invocation of a service could be requested directly by the user or by the orchestration engine that implements the orchestration schema defined by the designer for the process at the abstract service level and specified using BPEL4WS. Given an invocation, the Concrete Service Invocator is responsible of finding a concrete service that implements the abstract specification. The coordinator role is essential when alternative services have to be evaluated before invocation, and negotiation mechanisms have to be appropriately selected.

3   Coordination for the Negotiation of a Single Concrete Service

Establishing a negotiation between services involves many aspects, like contacting negotiating parties or results notification, that can be considered typical problems of service coordination. Dealing with web services the problem of coordination in a dynamic environment has been tackled from two sides [1], *horizontal* and *vertical coordination*. In our context, the problem of managing the negotiation is seen as a problem of coordination of different services. Thus, the different notions of coordination have to be adapted to our negotiation context:

- **Horizontal Coordination**: it manages negotiation and it involves contacting negotiating parties, message brokering and negotiation results notification to parties and to the service invocator in order to invoke the right service with the right quality attributes established by the negotiation process. This form of coordination is called horizontal because it is common to the entire set of negotiation protocols the architecture is able to deal with;

---

[1] The project web site is available at http://www.mais-project.it

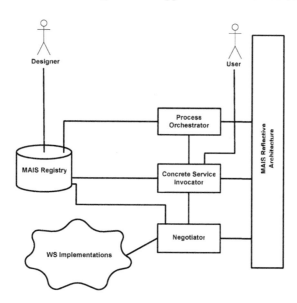

Fig. 1. The MAIS architecture for service provisioning

– **Vertical Coordination**: is the problem of specifying web services compliant descriptions of negotiating protocols and messages exchanged between parties that the platform is able to support. Examples found in literature, like [12], describe only this aspect of negotiation when dealing with web services architectures.

Besides the characterization of coordination, we describe a negotiation framework for concrete service invocation and how to refer to it while describing a process at abstract services level with BPEL4WS. Our effort is to use, where available, existing standards and specifications, without trying to introduce new languages every time we are facing problems in dealing with negotiation issues.

In our scenario, services can register on the MAIS registry by publishing, besides other information, the negotiation protocols supported. For example, many users publish on the registry different theater tickets booking services that are able to participate in an English auction for last minute tickets sold by a theater ticket service. When the theater ticket service is offering last-minute ticket, users can specify that their published services for ticket booking will participate in the auction. The Negotiator module of Figure 1 is responsible of implementing and managing the auction between the theater ticket service and users' services and of notifying results to auction participants and to the concrete service invocator that will invoke the theater ticket service in order to buy tickets for the participant who won the auction at the price established by the negotiation.

Following the scenario, we can identify three phases in the service negotiation:

1. **Matchmaking**: discovering services that will be involved in the negotiation process;
2. **Negotiation**: horizontal coordination of services involved in the negotiation process and vertical coordination referred to a particular negotiation protocol, i.e., auction, bilateral or multiparty negotiation;
3. **Results Notification**: to negotiation participants and to the service invocator.

3.1   Matchmaking for Negotiation

In order to be discovered by the service invocator when a negotiation process is needed or requested, services have to publish information about their negotiation capabilities. Policies are the most natural way to represent services' requirements, features or capabilities and, thus, we do not think that services have to implement a particular *portType* for every protocol supported, but supported negotiation protocols will be described as policies attached to the service, by means of WS-Policy [6]. Every policy will refer to a particular service that is responsible of handling different negotiation protocols (*handler services*).

Thus, policies defined in a service specification associate a service with a handler of different negotiation protocols that exports a *portType* for every protocol supported. The definition of a single policy is based on a schema of the information needed to describe negotiation protocols. An English auction protocol, for instance, is described by the maximum and minimum number of participants, the maximum and minimum length of the auction and the schema of messages used to post an offer in the auction. Figure 2 describes the elements required for matchmaking in a web service environment when negotiation capabilities are considered.

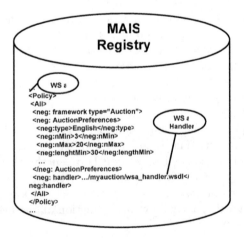

Fig. 2. Matchmaking for service selection

3.2   Horizontal and Vertical Coordination for Negotiation

In describing phase 2 we are facing with the situation where the service invocator has discovered all the services that will participate in the negotiation process
(i.e., the participants), and it has forwarded this information to the Negotiator with a reference to the negotiation protocol that has to be followed in the
negotiation process. The Negotiator acts as the unique coordinator of services
discovered and the coordination is described by means of WS-Coordination [8].
The UML sequence diagram of Figure 3 shows messages exchanged in the horizontal coordination phase.

The Negotiator "activates" services by sending to everyone involved a coordination context. The coordination context is a data structure that contains
the reference to the coordinator's *RegistrationPortType* and information about
the protocol that will be used in the negotiation process. Nevertheless, the coordination context is used to mark messages belonging to the same conversation
(i.e., negotiation process) because the Negotiator can be coordinating, at the
same time, different negotiations, for example an auction for theater tickets and
a direct negotiation of a user for purchasing a flight ticket.

After receiving the coordination context, each service registers to the coordinator specifying his role in the conversation and the portType that exports
the conversation operations; the coordinator replies with a reference to the port-
Type that exports the conversation coordination operations. For instance, in
the auction scenario, after receiving the coordination context, every service involved will register to the Negotiator specifying a portType, i.e., *AuctionPartic-
ipantPortType* for a participant. The Negotiator replies with a reference to his
AuctionCoordinatorPortType that manages the auction protocol.

The activation and registration phases refer to the horizontal coordination
of the negotiation protocol. The next step, in fact, is the exchange of protocol
specific messages, i.e., offers in the auction by services involved. In the management of the protocol specific interactions between handler services the Negotiator
assumes two roles:

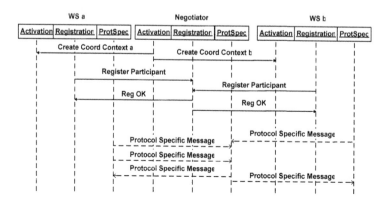

Fig. 3. Coordination for service selection

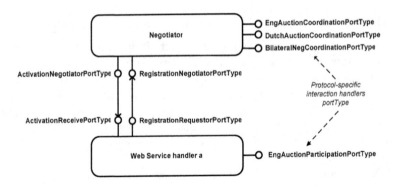

Fig. 4. PortType exported by the handler and the Negotiator

- **Negotiation Broker**: referring to horizontal coordination, the Negotiator delivers protocol specific messages to proper receivers and notifies negotiation results;
- **Protocol-compliance checker**: the Negotiator checks if conversations between services are compliant to the protocol specified in the coordination context; in an English auction, for instance, a participant is not allowed to post two consequent bids.

The description of each negotiation protocol and, thus, of the structure of the messages exchanged in each protocol supported could be made using the proposal of WS-Negotiation [12] or extending the constructs provided by WS-Transaction [7]. The Negotiator, see Figure 4, exports a different portType for every negotiation protocol supported (i.e., different kinds of auctions, bilateral or multiparty negotiation) and each service that will handle the negotiation exports at least a single portType for the negotiation protocol specified in the policy of the correspondent service.

3.3   Results Notification

After the exchange of protocol specific messages, the process turns again to horizontal coordination: results notification, to services involved and the Concrete Service Invocator, is the last set of interaction common to the entire set of negotiation protocols supported by the platform. Notified by the Negotiator, the service invocator is responsible of invoking a particular service with quality attributes defined by the outcomes of the negotiation process just terminated.

4   An Example of Protocol Definition

In order to act as a protocol-compliance checker, the Negotiator needs a web service compliant definition of negotiation protocols supported by the platform. The Negotiator uses this meta-definition to define the rules of the protocols and to define fault handlers for managing exception generated by a non compliant behaviour of services involved. Since all the executable process specifications in the MAIS framework are made using BPEL4WS, we chose it for the definition of *meta-protocols* for negotiation.

The following example is related to the definition of an English auction negotiation protocol. A timeout is associated with the auction, services involved can place bids until the timeout expires, the highest bid wins the auction; a bid is described by a sender id and an amount that he wants to pay. The Negotiator receives offers and notify results to the participants and to the service invocator. The compliance rule that the Negotiator has to check every time a new offer is submitted is that one service is not allowed to post two consequent offers. When, in fact, the last bidder is the same as the current bidder, the Negotiator throws an exception to signal the a non compliant behaviour; the exception handler, in this case, will be the refusal of the bid submitted.

```
<process name="auctionMetaSpec"
   ...
  <partnerLinks>
      <partnerLink name="buying"/>
  </partnerLinks>
  <variables>
      <variable name="lastBidder" type="xsd:string"/>
      <variable name="currBidder" type="xsd:string"/>
      <variable name="tmpSender" type="xsd:string"/>
      <variable name="tmpAmount" type="xsd:string"/>
      <variable name="winnerAmount" type="xsd:string"/>
      <variable name="winnerID" type="xsd:string"/>
   ...
  </variables>
  <faultHandlers>
   ...
  </faultHandlers>
  <sequence>
  <while condition="t < TIMEOUT">
  <receive partnerLink="bidder"
   portType = "bidPT"
   operation = "sendBid"
   variable = "offer"
  </receive>
  <assign>
  <copy>
      <from variable="offer" part="sender"/>
      <to variable="currBidder"/>
  </copy>
  </assign>
  <switch>
  <case condition="lastBidder=currBidder">
      <throw faultName="invalidBid" faltVariable="..."/>
  </case>
  </switch>
  <assign>
  <copy>
      <from variable="offer" part="sender"/>
      <to variable="tmpSender"/>
  </copy>
  <copy>
      <from variable="offer" part="amount"/>
      <to variable="tmpAmount"/>
  </copy>
  </assign>
  <switch>
  <case condition="tmpAmount>winAmount">
      <assign>
```

```
        <copy>
        <from variable=tmpSender/>
        <to variable="winnerID"/>
        </copy>
        <copy>
        <from variable="tmpAmount"/>
        <to variable="winnerAmount"/>
    </copy>
    </assign>
    </case>
    </switch>
    <assign>
    <copy>
        <from variable="offer" part="sender"/>
        <to variable="lastBidder"/>
    </copy>
    </assign>
</while>
        ...results notification...
    </sequence>
</process>
```

The process specified in the example is considered a meta-protocol because it does not refer to a particular abstract or concrete service. Roles in the negotiation protocol are specified only by names in the *partnerLinks* element, the Negotiator obtains a specification of the actual negotiation protocol enriching the meta-specification with the information provided by participants in the horizontal coordination phase. This mechanisms is similar to the already considered difference between abstract services and concrete implementations: in the meta specification *partnerLinks* are defined only from an abstract point of view, the coordination context gives a concrete binding to actual services involved in the negotiation process.The *bidPT* portType, for instance, will be every time substituted by the protocol specific interaction portType of the service that is placing the current bid.

5   Concluding Remarks and Future Work

In this paper we described a framework for managing negotiation in web service selection. The problem has been considered at the level of concrete service invocation and in the design of a process that involves the orchestration of different services. Future work will concern with the definition of languages for the specification of service coordination, the definition of different negotiation protocols and the specification of the need for negotiation in the process description.

The need for establishing web service parameters (e.g. [15]) for invocation using negotiation, in fact, is not only addressed by the problem of a concrete service invocation. The designer of the MAIS architecture of Figure 1 is responsible of designing process that will involve the orchestration of different services published on the registry. The process is always specified at abstract service level in a BPEL4WS file [2]. The process orchestration engine and the concrete service invocator will have to discover concrete services in the registry that will implement an instance of the process.

In this contest, we want to propose an extension to BPEL4WS to include the need for specification of the negotiation of service quality parameters. A typical scenario could be the one where, considering the process of setting up accommodation and travel for summer week-ends, a designer wants to specify that every Friday night a weather forecast service will be contacted and the negotiation issues are related to the price established for the service provisioning: in the first week-end the process hosts a buy side auction, where different forecast services will participate, and twice in a month the process will re-negotiate the terms of contract with the service that won the starting auction in an automated bilateral negotiation process. The aim of our extension is to provide mechanisms, inside BPEL4WS, for specifying different needs for negotiation in a process design at abstract level of service definition.

Acknowledgments

This work has been partially supported by the Italian MIUR-FIRB Project MAIS.

References

1. G. Alonso, F. Casati, H. Kuno, and V. Machirayu. *Web Services: Concepts, Architectures and Applications.* Springer-Verlag, Heidelberg, New York, 2004.
2. L. Baresi, D. Bianchini, V. D. Antonellis, M. G. Fugini, B. Pernici, and P. Plebani. Context-aware composition of e-services. In *Proceedings of the 4th VLDB Workshop on Technologies for E-Services TES'03*, Berlin, Germany, September 2003.
3. B. Benatallah, Q. Sheng, and M. Dumas. The self-serv environment for web services composition. *IEEE Internet Computing*, 7(1):40–48, February 2003.
4. B. Benatallah, H. Skosgrud, and F. Casati. Abstracting and enforcing web service protocols. *Int. Journal of Cooperative Information Systems*, 15(6):1345–1363, December 2003.
5. F. Casati, E. Shan, U. Dayal, and M.-C. Shan. Business-oriented management of web-services. *Communications of the ACM*, 46(10):55–60, October 2003.
6. D. Box et al. *Web Service Coordination (WS-Coordination).* IBM, Microsoft, BEA, http://www-106.ibm.com/developerworks/library/ws-polfram/, September 2003.
7. D. Box et al. *Web Services Transaction (WS-Transaction).* http://www-106.ibm.com/developerworks/webservices/library/ws-transpec/, August 2003.
8. F. Cabrera et al. *Web Service Coordination (WS-Coordination).* http://www-106.ibm.com/developerworks/library/ws-coor/, September 2003.
9. T. Andrews et al. *Business Process Execution Language for Web Services (BPEL4WS).* http://www-106.ibm.com/developerworks/library/ws-bpel/, May 2003.
10. P. Faratin, C. Sierra, and N. R. Jennings. Negotiation decision functions for autonomous agents. *Int. Journal of Robotics and Autonomous Systems*, 23(3-4):159–182, 1997.
11. M. Hu, H. Leung, and N. R. Jennings. A fuzzy-logic based bidding strategy for autonomous agents in continuous double auctions. *IEEE Transactions on Knowledge and Data Engineering*, 15(6):1345–1363, December 2003.
12. P. Hung, H. Li, and J.-J. Jeng. Ws-negotiation: An overview of research issues. In *Proceedings of the 37th Hawaii International Conference on System Sciences*, pages 84–89, Hawaii, USA, February 2004.

13. N. R. Jennings, P. Faratin, A. Lomuscio, S. Parsons, C. Sierra, and M. Wooldridge. Automated negotiation: Prospects, methods and challenges. *International Journal of Group Decision and Negotiation*, 10(2):199–210, 2001.

14. F. Lin and K. Chang. A multiagent framework for automated bargaining. *IEEE Intelligent Systems*, 16(4):41–47, August 2001.

15. C. Marchetti, B. Pernici, and P. Plebani. A quality model for multichannel adaptive information systems. In *Proceedings of the 13th World Wide Web Conference WWW04*, pages 49–55, New York, USA, May 2004.

16. S. Modafferi, A. Maurino, E. Mussi, and B. Pernici. A framework for complex e-service provisioning. In *Proceedings of the 1st IEEE International Conference on Services Computing SCC'04*, Shangai, China, September 2004.

17. The MAIS Team. Mais: Multichannel adaptive information systems. In *Proceedings of the 4th International Conference on Web Informaton Systems Engineering*, Rome, Italy, December 2003.

18. K. Binmore and N. Vulkan. Applying game theory to automated negotiation. In *DIMACS Workshop on Economics, Game Theory and the Internet*, Rutgers University, April 1997.

19. P. Faratin, C. Sierra, and N. R. Jennings. Negotiation decision functions for autonomous agents. *Int. Journal of Robotics and Autonomous Systems*, 23(3-4):159–182, 1997.

20. P. Faratin, C. Sierra, and N. R. Jennings. Using similarity criteria to make negotiation trade-offs. In *Proceedings of the 14th Int. Conference on Artificial Intelligence, AAAI'97*, Providence, Rhode Island, USA, July 1997.

21. M. Hu, H. Leung, and N. R. Jennings. A fuzzy-logic based bidding strategy for autonomous agents in continuous double auctions. *IEEE Transactions on Knowledge and Data Engineering*, 15(6):1345–1363, December 2003.

22. N. R. Jennings and S. Parsons. Negotiation through argumentation - a preliminary report. In *Proceedings of the 2nd Int. Conference on Multi-Agent Systems*, Kyoto, Japan, December 1996.

23. N. R. Jennings, P. Faratin, A. Lomuscio, S. Parsons, C. Sierra, and M. Wooldridge. Automated negotiation: Prospects, methods and challenges. *International Journal of Group Decision and Negotiation*, 10(2):199–210, 2001.

From Web Service Composition to Megaprogramming*

Cesare Pautasso and Gustavo Alonso

Department of Computer Science,
Swiss Federal Institute of Technology (ETHZ),
ETH Zentrum, 8092 Zürich, Switzerland
{pautasso,alonso}@inf.ethz.ch

Abstract. With the emergence of Web service technologies, it has become possible to use high level megaprogramming models and visual tools to easily build distributed systems using Web services as reusable components. However, when attempting to apply the Web service composition paradigm in practical settings, some limitations become apparent. First of all, all kinds of existing "legacy" components must be wrapped as Web services, incurring in additional development, maintenance, and unnecessary runtime overheads. Second, current implementations of Web service protocols guarantee interoperability at high runtime costs, which justifies the composition of only coarse-grained Web services. To address these limitations and support the composition of also fine-grained services, in this paper we generalize the notion of service by introducing an open service meta-model. This offers freedom of choice between different types of services, which also include, but are not limited to, Web services. As a consequence, we argue that service composition – defined at the level of service interfaces – should be orthogonal from the mechanisms and the protocols which are used to access the actual service implementations.

1 Introduction

Megaprogramming [23] was originally introduced to describe the large scale composition of megamodules, capturing the functionality of services provided by large, independent organizations. Megaprogramming prescribed a clear separation of the description of the externally accessible data structures and operations of a megamodule from the mechanisms used to interact with it. It also emphasized the importance of *mediation* between incompatibile megamodule descriptions.

Some of the existing languages for Web service composition (e.g. [5, 11]) do not yet completely fulfill the megaprogramming paradigm because the services

---

* Part of this work is supported by grants from the *Hasler Foundation* (DISC Project No. 1820) and the *Swiss Federal Office for Education and Science* (ADAPT, BBW Project No. 02.0254 / EU IST-2001-37126).

M.-C. Shan et al. (Eds.): TES 2004, LNCS 3324, pp. 39–53, 2005.

to be composed are all assumed to be of a single type: Web services. Clearly, when facing software integration problems at an Internet-wide scale, Web services seem to be the most appropriate tool [8]. However, for many other kinds of service integration scenarios, it would be an unnecessary restriction to assume that all services that are to be composed must all be Web service compliant. In fact, there are many existing, well established service access protocols (e.g., RMI, CORBA, JMS, HTTP) that should not necessarily be considered as out of date, when compared to Web services [1]. Furthermore, the *mediation* between incompatibile services turns out to be a very important requirement for successful integration projects. Thus, unless such "mediation services" themselves are encapsulated behind a Web service interface, it is not possible to efficiently address this important issue with current Web service composition languages [6].

In this paper, we show how we applied megaprogramming concepts to generalize Web service composition in the context of the JOpera project [15]. Web services can be considered as one *kind* of service, which is very useful, e.g., as it offers syntactical interoperability with remote services in a platform independent way [21, 22]. However, these benefits come at a price of a very high access overhead. This is justified for invoking coarse-grained services, for which the internal execution time dominates the overall invocation time. For other kinds of services, i.e., fine-grained services, which perform a small computation, or for local services, which are published within the same organization doing the service integration, it may be reasonable to employ other kinds of access mechanisms and protocols. This way, it is possible to choose the most appropriate service type in terms of the effort required to integrate it with others with the possibility of minimizing the corresponding invocation overhead. As it would be impossible to provide out-of-the-box support for all possible kinds of services, JOpera's service meta-model and the corresponding architecture can be extended to describe and interact with an open set of heterogeneous service access mechanisms.

This paper is structured as follows. In Section 2 we discuss related work in the context of Web service composition. In Section 3 we introduce JOpera's open service meta-model, followed in Section 4 by some examples on how to apply it to describe three (very) different kinds of services: Web services, Java snippets and legacy UNIX applications. In Section 5 we describe the relevant aspects of JOpera's architecture implementing the service meta-model. To give an indication of the difference between the cost of invoking coarse-grained Web services and fine-grained Java snippets we have included an overhead comparison in Section 6. In Section 7 we draw some conclusions.

2   Related Work

The need for supporting a variety of service access protocols is also recognized in the Web services community. To this end, the WSDL interface description standard supports an open-ended set of bindings. Therefore, a Web service, whose interface must be described using WSDL, does not necessarily need to be invoked using the relatively slow SOAP protocol if the client understands other (non standard) protocols which may offer better performance.

Currently, however, alternative protocols are not yet widely supported and as long as they are not standardized, using them would defeat the main point of the Web service vision, where everything should be standardized in order to achieve widespread interoperability [2].

Along these lines, the Web Services Invocation Framework (WSIF [9]) should be mentioned, as it provides this kind of access *transparency*. It allows to dynamically build clients to Web services described in WSDL, independent of the actual access mechanisms (e.g., SOAP) involved. As we will discuss in this paper, our service meta-model goes beyond that since it is not limited to services described with WSDL. Instead, it can be also applied to other interface description languages.

Moreover, in order to bridge the gap between the existing component heterogeneity and the uniform Web services standards, wrappers and interface adapters are still required to make the "legacy" types of components and protocols fit with the new standards. This approach introduces unnecessary execution overhead and shifts development and maintenance costs from the infrastructure to the end user [14]. Thus, we believe it is less expensive to build *once* a generic adapter to integrate a certain type of components into JOpera, instead of having to setup a different Web service wrapper for each of the service of that particular type that have to be integrated within a composite service.

Recently, to address the limitations of coarse-grained Web service composition, IBM and BEA systems proposed to extend the BPEL4WS [11] language with support for including Java snippets [10]. Although the need for such an extension was well argued, it remains unclear why, as opposed to Java, a .NET compliant language should not be chosen instead. Thus, a service composition technology which was originally tied to platform neutral Web services, becomes tangled into portability issues [20].

This problem originates from the confusion between the description of the composition and the description of the components. In our approach, we have chosen to keep a clear separation between the two. Thus, our visual service composition language [17] doesn't have to be modified to support new kinds of services, such as Java snippets, as this extension only affects the service meta-model.

3   An Open Service Meta-model

Before describing in detail the properties of some of the service types currently supported by JOpera, we introduce JOpera's open service meta-model. This way, we both motivate its flexibility and extensibility and summarize the information required to model and to access each type of service.

As shown in Figure 1, the interface of a service is defined in terms of a set of user-defined input and output parameters. This is the only information which is used in JOpera to define how the services are composed when drawing the data flow graph linking the parameters of different service interfaces [17].

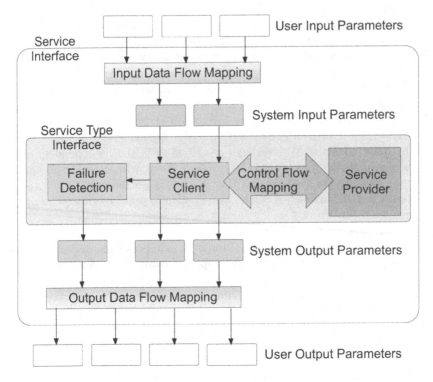

Fig. 1. Relationships between the various entities of the JOpera service meta-model

Thus, a service interface constitutes the minimal unit of composition. As a first approximation the mechanisms involved in the invocation of a specific type of service are kept completely transparent when modeling how to compose different service interfaces.

However, in order to support the actual invocation of a service, it is necessary to model additional information describing how to invoke its functionality and how to structure the data exchanged with it. Such information is abstracted into a *service type*. More precisely, when adding a new service type to JOpera's model it is necessary to define its interface (in terms of system parameters); design how to interact with it in terms of control and data flow; and devise a failure detection strategy.

Furthermore, the same service interface can be associated with multiple service types. This way, it becomes possible to choose between alternative service access mechanisms. On the one hand the service invocation can be dynamically adapted to the actual system configuration, whereby the most optimal mechanism is chosen depending on the current environment. On the other hand, if the invocation fails using one mechanism, another path can be attempted to access an alternative service provider.

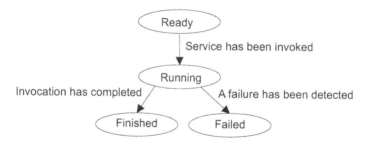

Fig. 2. Simple model of a service invocation

3.1   System Parameters

First of all, the interface of each service type is defined as a set of input ($[i]$) and output ($[o]$) parameters. These are called system parameters, to distinguish them from the user parameters, which are associated with the interface of the service. It is worth noting that user parameters depend on the specific application and therefore have nothing to do with the system parameters, which instead model the information required to access a particular type of service.

The input system parameters control the service invocation, as they identify the service and describe the information required to interact with the corresponding service provider. Their values are set at design time, when registering a new service with JOpera's component library.

The output system parameters model the raw results of the invocation as well as related metadata (e.g., status, performance profiling or debugging information). Their values are set after the invocation has completed and can be used to determine its outcome.

3.2   Control Flow

The transfer of control during one service invocation may involve different interaction patterns between the client and the service provider.

In the simplest case, the service is invoked synchronously, i.e., the client blocks until the results of the invocation are available. This case captures typical procedure-like invocations, e.g., a call to a local method, a remote procedure call, an HTTP request/response round.

However, other protocols involve the asynchronous (or event-based) interaction between client and service provider, based on the exchange of a pair of messages representing the starting of the invocation and the notification that it has completed. Following this protocol, the client does not block after sending the request to the service provider, although the invocation only completes after the client is notified with a response. Depending on the available mechanisms, the client may periodically poll the service provider for a response, or a notification message is pushed back from the service provider.

More sophisticated interactions with a service provider may involve the ability to abort, suspend and resume an ongoing invocation [19]. Likewise, it may be possible to retrieve partial results even before the whole invocation has completed [18].

In order to ensure the transparency of these different interaction patterns, we introduce a simple model of a single transfer of control between client and service provider in Figure 2. Using this model, a control flow mapping can be easily designed for the aforementioned synchronous and asynchronous cases. If necessary, the *Running* state can be extended to support other forms of interaction.

3.3   Data flow

From the point of view of transferring control, the interaction with different service types is not so difficult to model, as this amounts to describing the invocation of the service and the corresponding notification that the service's invocation has completed.

In our experience, a more difficult challenge lies in modeling the data to be exchanged with the service and in how to map JOpera's parameter based representation of its interface to the service's internal one. For some service types this can be relatively simple, at least from a syntactical perspective, where standards (e.g., SOAP) define how to format the input data and how to interpret the output data. In other cases, e.g., when integrating legacy UNIX applications, the problem is much more difficult and there is no general solution, i.e., the ad-hoc development of wrappers may be required.

In order to provide the necessary flexibility to integrate several different service types, in JOpera we follow a two step approach to address the problem of mapping user-level data parameters to the actual structure of the data understood by the service type.

The mapping between user (application) parameters and system (service type) parameters is specified once, when a new service component is registered with JOpera. This mapping can be derived automatically, e.g., by reading the WSDL description of a Web service.

The data flow mappings depicted in Figure 1 can be formally represented as a composition of two mappings (m_i, m_o) which are applied to fit the input and output parameters of a certain service call C to the given interface S. More precisely, the interface of a service contains a set of user-defined input ($[I]$) and output ($[O]$) parameters:

$$[O] = S([I])$$

Furthermore, a set of predefined service types C_t are available. These define the interface representation of the corresponding access mechanisms and invocation protocols in terms of input ($[i]$) and output ($[o]$) system parameters:

$$[o] = C_t([i])$$

In order to bind a service interface to an implementation of a given service type, it is necessary to provide the corresponding input and output mappings:

$$[i] = m_i([I])$$
$$[O] = m_o([o])$$

At runtime, these mappings are composed with the invocation of service of a given type as follows:

$$[O] = m_o(C_t(m_i(I))$$

Following such mapping, before a service can be invoked at runtime, the user input parameters are translated to its system input parameters. The main mechanism to model and perform this mapping (m_i) consists of using parameter placeholders, which identify one user input parameter and are replaced with its content when the mapping is evaluated. These placeholders follow the simple convention of including the name of a parameter between % characters [13].

The service is then invoked and the results are placed in the system output parameters corresponding to its type. The reverse mapping m_o from the system output parameters to the user-defined output parameters is applied. As opposed to the input mapping, where a relatively large number of user parameters are assigned to a small number of system parameters, in this case it is more complex to take the content of a few parameters, e.g., the output of a program or a Web page, and model how to extract the application dependent information. For data having a relatively well defined syntax, e.g., XML, it is possible to follow the convention of encoding parameter names as tags and insert their values between those tags [21].

In general, *ad-hoc wrappers* can be plugged into JOpera with the purpose of scraping the values of the output parameters from the arbitrarily formatted data produced by the service. Conversely, it is also possible to avoid breaking up the results of the invocation into output parameters and treat the result (e.g., in form of XML documents or other encodings) as a whole.

3.4   Failure Detection

Not only do service invocations finish; sometimes they fail. Depending on the type of service, failure detection may be based on different assumptions. For each type of service, it is important to devise a well-defined failure detection strategy, which determines the outcome of a service invocation. In case of failed invocations, a description of the problem involved can be stored in the corresponding system output parameters.

Furthermore, depending on the type of failure, different low-level error handling policies may be implemented. For example, the service invocation may be retried, if this option is supported by the underlying protocol. Thus, only unrecoverable failures occurring during the interaction with a particular service provider remain to be handled at the level of the service composition. In this case, exception handling constructs can be used to specify whether alternative (or compensating) services should be be invoked instead.

Table 1. Summary of the service types currently supported by JOpera

Service Type	Input and Output Data		Failure
WWW services			
Web Service	(SOAP) SOAP	SOAP	SOAP Fault
Web Server	(HTTP) CGI/URL	HTML	HTTP Error
Local services			
UNIX Application	(UNIX) CmdLine, Stdin Stdout		ExitCode, StdError
Java services			
Java Program	(JVM) CmdLine, Stdin Stdout		ExitCode, StdError
Java Snippet	(JAVA) Local Variables		Exception
Java Remote Method	(RMI) Method Parameters		Exception
Database services			
Database Query	(SQL) Parameters	XML	JDBC Error
XML services			
X-Path Query	(XPATH) XML	XML	X-Path Processor Error
Style Sheet Transformation	(XSL) Parameters	XML	XSLT Processor Error
System services			
JOpera Echo	(ECHO) XML	XML	XML Parser Error
JOpera Process	(OPERA)	Implicit Parameters and Failures	
Cluster/Grid computing services			
BioOpera [4]	(PEC) CmdLine	Stdout	ExitCode, StdError
Grid services [7]	(GLOBUS) SOAP	SOAP	SOAP Fault
Business process modeling services			
Workflow task	(WF) Text	Text	User Error

4   Examples

In this section we show how to apply our service meta-model to abstract the common features of different kinds of services. These represent three extreme cases: standard compliant Web services, fine-grained Java scripts and legacy UNIX applications.

Additionally, the current version of JOpera includes supports for many other kinds of services, modeling a Java remote method invocation (RMI), a job submitted to a batch scheduling system of a cluster of computers, an SQL query to be sent to a database, the asynchronous exchange of messages through a queuing system, a human activity, and an XSL style sheet transformation to be applied to some XML data packet [16]. In Table 1 we summarize the main properties of some of the service types to which we have applied JOpera's service meta-model.

4.1   Web Services

This first type of services models the latest form of standard compliant Web services, whose interface and location are described in a WSDL document [22] and which are remotely accessible through the SOAP protocol [21]. Web services offer the benefit of standard-based interoperability between heterogeneous programming languages and platforms. With this technology, the effort of building systems composed out of services distributed across the Internet is greatly reduced, at the price of a relative high runtime overhead due to the nature of the protocols involved. Thanks to these standards, it is possible to automatically import the service's WSDL description into JOpera's component library and use it to generate the corresponding service declarations automatically.

System Parameters. The invocation Web service is described by the following system input parameters: WSDL, with the URL used to locate the description of the service; service, operation, port, with the names of the WSDL elements used to identify the actual service, operation and port to be invoked; soapin, which contains the complete envelope of the SOAP request message to be sent when invoking the service. This includes both the header and the body of the SOAP request message. The response (or fault) message returned by a Web service is stored in the soapout system output parameter.

Data flow. The values of the user-provided input parameters are inserted in the SOAP request message using the previously described placeholder mechanism. In most cases, each input parameter corresponds to a SOAP message block. If necessary, JOpera escapes the content of the parameters so that it conforms to the required SOAP/XML encoding. The output parameters are filled by parsing the SOAP response message.

Failures. The invocation of a Web service may fail for several reasons: its WSDL description may be invalid; no response message from the service has been received after a certain timeout has expired; the service has responded with a soap fault message.

4.2   Java Snippets

This service type models the most efficient way of invoking Java code. By design, such code (or snippet) is embedded by the compiler into the code generated for a process. Thus, it can be invoked with minimal overhead. It can be very beneficial to use this kind of service to perform small computations [10]. Java snippets can be applied to perform data conversions, transforming the data in transit between incompatible services. Also, it gives a convenient syntax for the evaluation of complex conditional expressions. If the same computation would have to be invoked using a different mechanism (e.g., Web services), the overhead of the protocols involved would make it impractical to do so.

System Parameters. For Java snippets, there is only one system input parameter (script) which contains the Java code itself. If an error occurs, the exception system output parameter contains the message of the Java exception.

Data flow. There is a one to one correspondence between user defined parameters and the Java variables that can be implicitly used in the script. JOpera's compiler automatically declares Java variables for each input and output parameters. After the snippet has completed, the values assigned to the Java variables are copied into the corresponding output parameters.

Failures. JOpera detects a failure if a Java exception is raised and it is not caught during the execution of the script.

4.3   UNIX Applications

Another type of services, quite different from remote Web services, are commands to be executed in a shell of the local operating system. A shell command is typically used to provide a generic mechanism of invoking entire "legacy" applications. As long as these applications do not provide an explicit API, the command line may be the only viable mechanism to allow JOpera to interact with such applications and control their execution. In other words, this type of service is used to access the services provided by essentially any executable program, which is started by typing a command line at the prompt of the operating system shell.

System Parameters. As it is reflected by its system parameters (`command`, `stdin`, `stdout`, `stderr`), JOpera employs both the command line itself and pipe-based interprocess communication mechanisms in order to exchange data with the external program. Furthermore, the `retval` system output parameter contains the program exit code.

Data. The values of the user input parameters are transferred to the external program both using its `command` line and can also be copied onto its `stdin` system input parameter. If necessary, the `stdout` parameter can be parsed by a user-provided plugin to extract relevant information to be assigned to the user defined parameter.

Failures. JOpera interprets the value of the `retval` system parameter, which contains the exit code of the process as it is returned by the operating system, to distinguish between a successful execution (0) and a failed execution (non-0). In both cases, it also stores the program's standard error into the `stderr` parameter so that the user can gather useful debugging information.

5   Architecture

In order to support an open and heterogeneous set of service invocation mechanisms, JOpera's architecture uses plugins to extend the system's behavior at three different stages: service definition, service compilation and service invocation.

As shown in Figure 3, the Service Library Manager uses service import plugins to automatically import services described using other meta-models

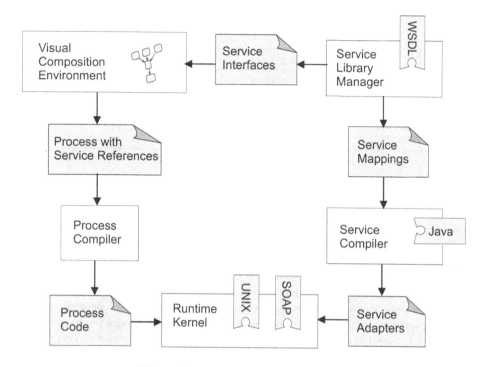

Fig. 3. JOpera plugin based architecture

(e.g., WSDL). Using the Visual Composition Environment, the developer may browse through the service library and select the service interfaces to be composed into processes [17]. During process compilation, all of the data flow mappings, which are part of the services referenced by a process are compiled into service adapters[1]. By default, the service compiler produces an efficient executable representation of the data flow mappings of a service. However, the service compiler can be extended with plug-ins corresponding to a specific type of service. For example, in case of Java snippets, the Java code entered as part of the aforementioned `script` parameter is injected into the resulting service adapter code, surrounded by the variable declarations corresponding to the user-defined parameters.

At run-time, the service invocation proceeds as depicted in Figure 1. The runtime kernel uses the compiled service adapters to perform the input and output data flow mappings, while the service is invoked through a kernel plugin. Such plugin uses the mechanisms and protocols specific to a certain service

---

[1] Although it is always possible to merge the code of the process with the service adapter code at compile-time, this would fix the binding between service interface and invocation adapter. Thus, in order to support late binding, the code of the process only contains references to services, which are resolved at the latest possible time.

Table 2. Service Invocation Mechanisms to be compared

Service Type	Description
JAVA	Java Snippet
UNIX	UNIX Application
SOAP/A11	Local Web Service using Axis 1.1 [3].
SOAP/A12	Local Web Service using Axis 1.2α.
SOAP/WS	Remote Web Service using Axis 1.1.

type (e.g., UNIX, SOAP) to interact with the service provider and perform the service invocation. Considering the service meta-model presented in Section 3, these plug-ins define the control flow mapping and the failure detection strategy for a given type of services and exchange information with the service adapters through system parameters. The kernel plugins are loaded on-demand, so that the system can be dynamically extended to deal with new types of services.

When adding support for a new type of service, a kernel plug-in is required. A compiler plugin is only necessary if the service adapter should perform some special processing before or after the invocation. A service import plugin can be added if it is possible to automatically generate JOpera service definitions from other interface description languages.

6   Overhead

Performance is one of the arguments behind the idea of providing support for invoking services of different service types. In order to give an indication of the overhead involved, we compare the time required by JOpera to invoke a remote Web service across the Internet with the time JOpera takes to perform a local Java method call, and – quantitatively – determine the cost (or the benefit) of preferring services of a certain type over another.

As listed in Table 2, in this performance comparison we use services of various types and several implementations of the corresponding kernel plugins.

More precisely, in this experiment we compare different access mechanism to the same "Temperature Conversion Service". We chose this service due to its trivial implementation, so that the execution cost is negligible when compared to the overhead of invoking it. Another reason to choose this service is that we found a remote implementation on the Internet at [12]. With it, it becomes possible to present an interesting comparison between the invocation overhead of local and remote Web services.

As shown in Figure 4, the most important result of this simple experiment is that the average service invocation overhead varies about three orders of magnitude (from about 1 millisecond to 2.31 seconds) depending on the service type.

The invocation of the Java snippet (JAVA) service offers an invocation overhead of significantly less than $1/100^{th}$ of a second, as the implementation of the

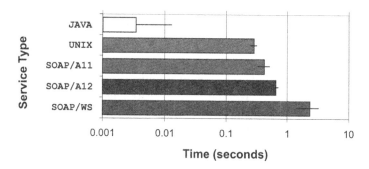

Fig. 4. Service Invocation Overhead for different service types

service is located within the same Java virtual machine where the JOpera kernel is running.

Invoking the UNIX application requires to spawn a child process through the local operating system, and this requires more time: about 0.28 seconds.

The average Web service invocation time is 0.42 seconds in case of a Web service deployed on the local area network, called using Axis version 1.1 (SOAP/A11). This time grows to 0.66 seconds using the latest version of Axis 1.2α (SOAP/A12). In case of the invocation of remote Web service with Axis 1.1 (SOAP/WS), the delay and jitter of the wide area network need to be discounted. This effect can be recognized both in the higher (2.31 seconds) average response time and in the very high standard deviation (0.9 seconds).

As expected, Web services are the most expensive service type in terms of the overhead involved. Given the current state of flux of the relevant standards and available implementations, the performance of the service invocation may be significantly affected by the choice of which libraries are used. Additionally, the location of the Web service also affects the overhead, as the cost of invoking the remote Web service shows.

Since this additional cost is due to the distributed nature of the service interaction, it should not be blamed on the Web services protocols, which – instead – are one of the few technologies currently enabling such type of distributed interaction. Nevertheless, such overhead should be paid only when necessary, i.e., to invoke remote services, while more efficient mechanisms should (and can) be used to access local services.

7   Conclusion

The main contribution of this paper lies in the idea that service composition should be orthogonal with respect to the types of components involved. By introducing a clear separation between the service *composition language* and the *service meta-model*, we are able to isolate the description of how to compose the services from how to invoke them. This approach is similar to megaprogramming [23], as it gives several conceptual and practical advantages. First of all, it is not necessary to extend the composition language if a new kind of service access mechanism has to be included, as this affects only the component model. Likewise, if it is possible to redefine the access mechanism (e.g., synchronous vs. asynchronous) to be employed without modifying the corresponding service interface, such modifications are completely transparent as far as the description of the composition is concerned. Such flexibility also leads to the possibility of doing optimizations since it becomes possible to choose the most efficient mechanisms and protocols to access both fine-grained and coarse-grained services. We are currently investigating several policies to autonomously select the optimal mechanism. This is much more difficult to accomplish if the services to be composed are restricted to only one type.

Finally, we believe that the possibility of choosing (wisely) between the use of Web Services or other kinds of services can be of great value, as the most appropriate type of service in terms of performance, security, reliability and convenience of use can be chosen.

References

1. G. Alonso. Myths around Web services. *Bulletin of the IEEE Technical Committee on Data Engineering*, 25(4):3–9, December 2002.
2. G. Alonso, F. Casati, H. Kuno, and V. Machiraju. *Web services: Concepts, Architectures and Applications*. Springer, November 2003.
3. Apache Software Foundation. *AXIS version 1.1.* http://xml.apache.org/axis.
4. W. Bausch, C. Pautasso, R. Schaeppi, and G. Alonso. BioOpera: Cluster-aware computing. In *Proceedings of the 2002 IEEE International Conference on Cluster Computing (CLUSTER 2002)*, pages 99–106, Chicago, IL, USA, 2002.
5. BPMI. *BPML: Business Process Modeling Language 1.0.* Business Process Management Initiative, Match 2001. http://www.bpmi.org.
6. C. Bussler. Semantic Web services: Reflections on Web Service Mediation and Composition. In *Proceedings of the Fourth International Conference on Web Information Systems Engineering (WISE 2003)*, pages 253–260, Roma, Italy, December 2003.
7. I. Foster, C. Kesselman, J. Nick, and S. Tuecke. The Physiology of the Grid: An Open Grid Services Architecture for Distributed Systems Integration. Technical report, Service Infrastructure Workgroup, Global Grid Forum, 2002. http://www.globus.org/research/papers/ogsa.pdf.
8. K. Gottschalk, S. Graham, H. Kreger, and J. Snell. Introduction to Web services architecture. *IBM Systems Journal*, 41(2):170–177, 2002.
9. IBM and Apache Foundation. *Web Service Invocation Framework (WSIF)*, 2003. http://ws.apache.org/wsif/.
10. IBM and BEA Systems. *BPELJ: BPEL for Java technology*, March 2004. http://www-106.ibm.com/developerworks/webservices/library/ws-bpelj/.
11. IBM, Microsoft, and BEA Systems. *Business Process Execution Language for Web services (BPEL4WS) 1.0*, August 2002. http://www.ibm.com/developerworks/library/ws-bpel.
12. C. Jensen. *Temperature Conversion Service*. http://developerdays.com/cgi-bin/tempconverter.exe/wsdl/ITempConverter.

13. F. Leymann and D. Roller. Business Process Management With FlowMark. In *Proceedings of the 39th IEEE Computer Society International Conference (CompCon '94)*, pages 230–234, February 1994.

14. J. Oberleitner and S. Dustdar. Constructing Web services out of Generic Component Compositions. In *Proceedings of the International Conference on Web services (ICWS-Europe 2003)*, pages 37–48, Erfurt, Germany, 2003.

15. C. Pautasso. JOpera: Process Support for Web services.
http://www.iks.ethz.ch/jopera/download.

16. C. Pautasso. *A Flexible System for Visual Service Composition.* PhD thesis, Diss. ETH Nr. 15608, July 2004.

17. C. Pautasso and G. Alonso. Visual Composition of Web Services. In *Proceedings of the 2003 IEEE International Symposium on Human-Centric Computing Languages and Environments (HCC 2003)*, pages 92–99, Auckland, New Zealand, 2003.

18. N. Sample, D. Beringer, and G. Wiederhold. A Comprehensive Model for Arbitrary Result Extraction. In *Proceedings of the 2002 ACM symposium on Applied computing (SAC 2002)*, pages 314–321, Madrid, Spain, 2002.

19. H. Schuster, S. Jablonski, P. Heinl, and C. Bussler. A General Framework for the Execution of Heterogeneous Programs in Workflow Management Systems. In *Proceedings of the 1st IFCIS International Conference on Cooperative Information Systems (CoopIS'96)*, pages 104–113, Los Alamitos, CA, 1996. IEEE Computer Society Press.

20. H. Smith. *Enough is enough in the field of BPM: We don't need BPELJ: BPML semantics are just fine,* April 2004.
http://www.bpm3.com/bpelj/BPELJ-Enough-Is-Enough.pdf.

21. W3C. *Simple Object Access Protocol (SOAP) 1.1,* 2000.
http://www.w3.org/TR/SOAP.

22. W3C. *Web services Definition Language (WSDL) 1.1,* 2001.
http://www.w3.org/TR/wsdl.

23. G. Wiederhold, P. Wegner, and S. Ceri. Towards Megaprogramming: A Paradigm for Component-Based Programming. *Communications of the ACM*, 35(11):89–99, 1992.

Using Process Algebra for Web Services: Early Results and Perspectives

Lucas Bordeaux and Gwen Salaün

DIS - Università di Roma "La Sapienza",
Via Salaria 113, 00198 Roma, Italia
{bordeaux,salaun}@dis.uniroma1.it

Abstract. Web services are computational entities distributed on the web whose goal is to cooperate in order to work out simple or complex tasks. In this paper, we advocate the use of process algebra as an abstract and formal description formalism to tackle several issues raised in the context of web services. Abstract processes are helpful to describe services at different levels of expressiveness depending on the goal at hand and to compose them in order to build more complicated services. A great interest of using process algebra is that formal reasoning is made possible at any time and for many purposes (*e.g.* composition correctness) thanks to the existence of state-of-the-art tools. Abstract descriptions may also be used as a first step to develop certified web services following a well-defined method. We discuss all these ideas in this paper, reinforcing them with simple examples.

1 Introduction

Web Services (WSs) emerged recently and are a promising way to develop applications through the internet. WSs are distributed, independent pieces of code (one might also call them *processes* due to their behavioural foundation) which communicate with each other through the exchange of messages. A central question in WS engineering is to make a number of processes interact together to work out a given task. WSs raise many theoretical and practical issues which are part of on-going research. Some well-known open problems related to WSs are to specify them in an adequate, formally defined and expressive enough language, to compose them (possibly automatically), to discover them through the web, to ensure their correctness, etc.

It is becoming well-admitted that the use of formal methods is worthy as an abstract way to deal with WSs and then to tackle several issues raised in WSs [4]. In this direction, process algebras (PAs) have been used recently at different levels [24, 4, 29], particularly because they focus on the description of behaviours, consequently they are appropriate to specify the exchange of messages between WSs. Additionally, many process calculi (CCS, TCSP, LOTOS, π-calculus, Promela, etc) and accompanying tools (CWB-NC, CADP, MWB, SPIN, etc) exist, which offer a wide panel of expressiveness to deal with valuable issues (*e.g.* formal reasoning) in the WS area.

M.-C. Shan et al. (Eds.): TES 2004, LNCS 3324, pp. 54–68, 2005.
© Springer-Verlag Berlin Heidelberg 2005

The goal of this paper is to itemize and discuss some of these issues for which PA is helpful. Therefore, after an introduction of process algebra in Section 2, we will describe in Section 3 how PA and its tools may be used (i) to reason on WSs to ensure properties of interest, (ii) to develop WSs from abstract descriptions following a well-defined process, (iii) to compose WSs which could be an adequate solution to the choreography issue and (iv) to orchestrate them. Section 4 compares the use of PA to other description languages (especially based on transition systems) and Section 5 ends with concluding remarks.

2 What is a Process Algebra?

A PA is an abstract language to specify concurrent processes. A PA is based on the notion of *action* corresponding to a local evolution of a process. Actions are above all the way to express interactions (emissions and receptions) between two (or more) processes. Basic operators to describe dynamic behaviours are the sequence of actions, the nondeterministic choice and the parallel composition (denoted respectively ., + and | in the following). Furthermore, PA introduces other constructs and uses different communication models (*e.g.* binary *vs* multi-party communication, synchronous *vs* asynchronous). These calculi are formalised axiomatically with algebraic laws which can be used to verify term equivalences, and/or operationally using a semantics based on Labelled Transition Systems. These formalisms are most of the time tool-equipped, enabling one to simulate possible evolutions of processes, to check properties (*e.g.* to ensure that a bad situation never happens) and to verify equivalences.

For illustration purposes, let us introduce a very simple piece of behaviour written in a subset of CCS and its possible evolutions computed from its operational meaning (Fig. 1). Synchronizations occur on complementary actions[1] (one emission and one reception, respectively denoted $'a$ and a for example) and the result of the synchronization is a τ action. A terminated process is written 0 (*"do nothing"*). In the execution tree (representing all possible executions of a system) below, two processes are composed in parallel and may interact first either synchronizing on the *error* action (left hand side arrow) or synchronizing on the *req* action (right hand side arrow). Afterwards, for the right part, the first process evolves along a local computation and may synchronize with the other one on an *ok* or *refusal* action.

Now, we aim at surveying some languages that, in our opinion, are of interest for WSs. All these calculi are as many possible choices one may prefer depending on the aspects of services to be captured. Our goal is not to be exhaustive but to let the reader know about existing material.

---

[1] In the basic CCS, synchronizations are explicitly defined using the restriction operator \. For the sake of readability and comprehension, we consider here that they occur between actions gathered in a synchronization set as illustrated in the example.

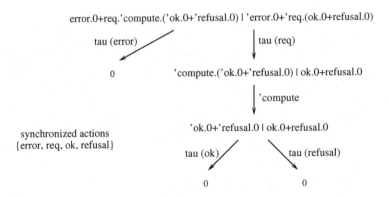

Fig. 1. Possible evolutions of a simple CCS process

Data Descriptions. With regards to this aspect, let us introduce LOTOS. It is an ISO specification language [16] which combines algebraic abstract types (data and operations) and dynamic processes (inspired from CCS and CSP). The point of interest is its expressiveness, particularly to describe rich data structures (*e.g.* integers, list of integers, sets, etc). Each *sort* (or datatype) defines a set of *operations* with arity and typing (the whole is called *signature*). A subset of these operations, the *constructors*, are used to create all the elements of the sort. *Terms* are obtained from all the correct operation compositions. *Axioms* are first-order logic formulas which define the meaning of each operation appearing in the signature.

Regarding the description of behaviours, LOTOS is made up of the classical basic constructs extended with more advanced ones (interruption of a process by another one, different parallel and sequential composition operators, etc). At the syntactic level, data terms are directly integrated within the dynamic constructs (local data, guards, value passing, exits). CADP[2] is a toolbox supporting developments based on LOTOS specifications. It proposes a wide panel of functionalities ranging from interactive simulation to formal verification techniques (minimization, bisimulation, proofs of temporal properties, compositional verification, etc).

```
process Store[request, ok, nok]   (* process parametrized by 3 channels  *)
         (s:Stock): exit :=       (* and a variable 's'                  *)

  request?id:Nat?q:Nat;           (* receive two numbers 'id' and 'q'    *)
  (                               (* from a channel named 'request' ;    *)
    [isAvailable(id,q,s)] ->      (* if we have q items in stock then    *)
      ok;                         (* emit ok and update the stock        *)
      Store[request, ok, nok] (decrease(id,q,s))
    [] (* OR *)
      ... (* other choices omitted *)
```

---

[2] http://www.inrialpes.fr/vasy/cadp/

The piece of LOTOS specification below describes the management of a stock of products in a store. We especially focus on the reception of a request characterized by the identifier of a product and a quantity. The operation isAvailable checks the availability of such a quantity in the local stock. In case of positive answer, an acceptance is communicated along the ok channel.

Asynchronous Communication. An example of a calculus based on asynchronous communication is Promela [14]. Promela specifications consist of processes, message channels, and variables. Promela has a C-like notation. This modelling language is particularly well-known because it is the input language of SPIN which is a state-of-the-art tool for analyzing the logical consistency of distributed systems (esp. based on LTL model checking). Basic datatypes (*e.g.* integers, boolean) may be handled and one advanced construct (array) is available as well to describe data. The language allows for the dynamic creation of concurrent processes. Communication via message channels can be defined to be synchronous (*i.e.* rendez-vous), or asynchronous (*i.e.* buffered).

Let us give a short piece of Promela code illustrating the language and the need for asynchrony (and accordingly reasoning means on such descriptions). The behaviour below shows a store receiving requests from a client and answering positively or not depending on a value transmitted by the client. The interesting point is that the store cannot start replying to requests before it receives a start authorization. However, requests may be emitted from clients and then buffered until the store is able to treat them.

```
proctype store(chan start, req, reply)   /* 3 channel parameters    */
{
    start?x;                          /* receive x on channel 'start'  */
    do                                /* loop forever:                 */
    :: req?y;                         /* receive value y on chan 'req' */
        if                            /* test (whether y > 5)          */
        :: (y > 5)  -> reply!1        /* if so, send 1 on reply channel */
        :: (y <= 5) -> reply!0        /* otherwise send 0 on it        */
        fi
    od
}
```

Mobility. The π-calculus [28] is a simple process algebra (based on CCS) dedicated to mobility and reconfiguration. Such functionalities are expressed using name passing: one process may receive a name attached to an action that it uses afterwards dynamically as a new communication port. Dynamic instantiations are possible in this calculus using the spawning (or replication) operator. The Mobility Workbench (MWB) is a prototype making it possible to prove temporal properties and to check equivalences between processes written using the π-calculus. Let us remark that other languages and tools (like Promela/SPIN) allow a limited form of name passing.

For illustration purposes, we introduce a part of a bank service. A bank receives first a request from a client and the port on which the client wants to send his private code. Consequently, the *port* message is used after reception as a classical communication channel. The bank emits on this channel too the requested information. Every time the bank receives a request from a client, it replicates itself in order to be ready to accept a new request.

```
agent Bank(request) =    (* process parametrized by name 'request'   *)
    request(port).(       (* receive name 'port' on channel 'request'  *)
        Bank(request)     (* create in parallel a new main process and *)
                          (* one which receives the code and emits info *)
    | port(code).(^info)('port<info>.0))
```

3 Tackling Some WS Issues Using PA

In this section, we introduce and describe for what issues PAs are helpful. First of all, we emphasize that abstract processes may be used at different levels of description depending on what we want to represent, public (interfaces) or private (execution details) aspects of its behaviour with more or less details regarding the usefulness of such descriptions. Abstract descriptions are achieved in two ways: (i) we can develop WSs considering the formal and verified specification as a starting point (design stage); (ii) in the other direction, we can abstract interacting executable services (described in an XML-based style or with a classical programming language) to a description in PA (reverse engineering).

3.1 Formal Reasoning

The major interest of using abstract languages grounded on a clear semantics is that tools can be used to check that a system matches its requirements and works properly. *Model checking* [7] is the preferred technique at this level (especially compared to the *theorem proving* alternative [13]) because it works with automata-based models (underlying models of process-algebraic notations), and proofs are completely automated (press-button technology). Specifically, these tools can help (i) checking that two processes are in some precise sense *equivalent* – one process is typically a very abstract one expressing the specification of the problem, while the other is closer to the implementation level; (ii) checking that a process verifies desirable *properties* – e.g. the property that the system will never reach some unexpected state.

Intuitively, two processes or services are considered to be equivalent if they are *indistinguishable* from the viewpoint of an external observer interacting with them. This notion has been formally defined in the PA community [25], and several notions of equivalence have been proposed: trace equivalence, observational equivalence, strong bisimulation are the most relevant ones within the context of WS. In [29], we show how such equivalence notions can be used to check compatibility between services and then to ensure the correctness of a composition.

The properties of interest in concurrent systems typically involve reasoning on the possible scenarios that the system can go through. An established formalism for expressing such properties is given by *temporal logics* like CTL⋆ [20]. These logics present constructs allowing to state in a formal way that, for instance, all scenarios will respect some property at every step, or that some particular event will eventually happen, and so on. The most noticeable kinds of properties one may want to ensure are:

- *safety properties*, which state that an undesirable situation will never arise;
- *liveness properties*, which state that something good must happen.

With regards to the reasoning issue, works have been dedicated to verifying WS descriptions to ensure a correct execution and properties of interest [29, 10, 8, 27, 9, 26]. Summarizing the approaches proposed so far, they verify some properties of cooperating WSs described using XML-based languages (DAML-S, WSFL, WSDL, BPEL4WS, WSCI). Accordingly, they abstract their representation and ensure some properties using ad-hoc or existing tools.

Reasoning abilities are strongly related to the visibility level. In case of reverse engineering approaches, if processes are viewed as white or glass boxes, reasoning on interacting processes is possible. However, processes are usually black boxes (internal details are not unveiled to users) and the verification is restricted to reason on visible interfaces or traces (execution results). On the other hand, in case of development of certified WSs (described further in this paper), reasoning steps can be straightforwardly carried out. Depending on the description model chosen for WSs (and on the level of abstraction), an adequate language and tool may be preferred. Another criterion which may influence such a choice is the kind of checking to be performed.

Last but not least, we stress that a checking stage may be helpful for several reasons and at different levels (as introduced above). In the next subsections, even if we focus on other issues (development, composition, choreography), reasoning needs are of strong interest for all these problems.

3.2   Developing Certified WSs

Designing and developing certified and executable WSs is a promising direction and very few proposals are dealing with such a question in a formal way. A lot of work remains to be done to have at one's disposal a smooth and formal process (known as refinement) enabling one to develop (correct) running WSs from abstract and validated descriptions.

At this level, we especially experimented the use of PA to develop executable and certified WSs from such abstract descriptions [30, 6]. They are especially worthy as a first description step because they enable one to analyze the problem at hand, to clarify some points, to sketch a (first) solution using an abstract language (therefore dealing only with essential concerns), to have at one's disposal a formal description of one or more services-to-be, to use existing reasoning tools to verify and ensure some temporal properties (safety, liveness and fairness properties), to encode processes into executable services (we illustrate in the following with implementations developed using the BPEL standard), and finally to execute them.

The link between the abstract level and the concrete one is formalised through systematic guidelines which make it possible to translate such abstract processes into WSDL interfaces and BPEL processes. Depending on the process algebra (and particularly on its expressiveness) used in the initial step, running BPEL services or just skeletons of code (to be complemented) may directly be obtained. We emphasize that such an approach is valuable for many applications in e-commerce: auction bargaining, on-line sales, banking systems, trip organizations, etc.

We illustrate our approach on a simple example; the reader interested in the development of more complicated WSs following this approach may refer to [30, 6]. The service to be developed aims at storing some private information which can be accessed by clients. Each client sends its identifier and receives back the corresponding stored information. For simplification reasons, we represent all the managed data (identifiers, stored information) as natural numbers.

Since this service has to manage (possibly more complex) data, we specify the behaviour of our service using the LOTOS calculus. The process AccessServ below receives first an identifier along the request channel, tests whether this identifier belongs to its local list 1, and depending on the guard value sends the corresponding piece of information (retrieved using the extractI operation) back or emits the error code 0.

```
process AccessServ [request,reply](l:Info): exit :=

    request?id:Nat; (
        [isIn(id,l)]         ->        (* if we have the id           *)
            reply!extractI(id,l);exit  (* return its associated info *)
        []                             (* OR                          *)
        [not(isIn(id,l))] ->           (* if we do not                *)
            reply!0;exit               (* send value 0                *)
    )
endproc
```

This abstract description has been validated using the CADP toolbox, even if the simplicity of the case at hand does not require advanced reasoning steps to check the correct processing of the service. We note that the process is not recursive because it corresponds to the behaviour of one transaction which would be instantiated as many times as needed.

Let us sketch the necessary guidelines to translate such a basic process in WSDL and BPEL. Firstly, LOTOS emissions and receptions (gates, variables and types) are encoded in WSDL using messages which are completely characterized by the message, the port type, the operation and the partner link type. The list of couples is encoded as a simple database containing one table with two fields *(identifier, info)*. A Java class has been encoded and contains methods to access the database using SQL queries, *e.g.* to test the presence of an identifier within one couple. The reception and the emission are translated as a *receive* activity followed by a *reply* one. Parameters of the messages are adequately stated using the *assign* activity before sending. The choice is composed of two branches in which guards are used to condition the firing, thereby a *switch* activity is employed to translate this LOTOS behaviour, and each branch is implemented using a *case* activity whose guard corresponds to querying the database. We end with the skeleton of the BPEL code describing interactions between the access service and a client. Experimentations have been carried out using Oracle BPEL Process Manager 2.0[3] that enables one to design, deploy and manage BPEL processes.

```
<sequence name="main">
  <receive name="receiveInput" partnerLink="client"
           portType="tns:AccessServ" operation="getNumber"
           variable="request" createInstance="yes"/>
      ...
      <!-- connecting to the db to check the presence of the couple -->
      <bpelx:exec language="java" version="1.4">
           ...
              //open the connection with the DB
              /* verify if the couple is present in the list */
              //close the connection with the DB
           ...
      </bpelx:exec>
      <switch>
          <case condition="bpws:getVariableData('available') = true()">
              <sequence>
                  <bpelx:exec language="java" version="1.4">
                      ...
                      /* retrieve the second value from the couple  */
                      ...
                  </bpelx:exec>
                  <assign> <copy> ... </copy> </assign>
                  <reply name="replyOutput" partnerLink="client" ... />
              </sequence>
          </case>
          <otherwise>
              <sequence>
                  <assign> ... </assign>
                  <reply name="replyOutput" partnerLink="client" ... />
              </sequence>
          </otherwise>
      </switch>
</sequence>
```

3.3   Choreography

E-business applications are developed from existing material rather than built from scratch, and an ever-increasing part of the programming activity is devoted to producing complex software by reusing and gluing together existing components. *Composition* is the problem of finding the right way to put together a judicious set of WSs in order to solve a precise task. Regarding this general definition, *choreography* aims at defining a global model for services describing

---

[3] http://www.collaxa.com/

precisely their interactions. On the other hand, *orchestration* is more dedicated to coordinating WSs involved in a given composition for example. Herein, we advocate the use of PA for either notion related to the composition question, particularly because PAs are compositional languages. Accordingly, it is possible to build complex WSs composing basic ones, and this can be carried out using every construct pertaining to any PA.

Let us start focusing on the choreography issue. A central problem in choreography is to find out a judicious expressiveness level for WS public interfaces and to use them for composition purposes in a next step. As made explicit by the W3C choreography working group [33] it is now accepted that, in a near future, the interface of WS should evolve and that a description of their *observable behaviour* should be provided in addition to their sole WSDL interface. Additionally, these interfaces have to be complemented by a precise way to define how web services interact together. These descriptions will be based on an XML-based standard like WSCI [32], BPEL [1] or WS-CDL [31]. PA may be viewed as solution for this issue at an abstract description level [5, 29] but may also be considered at a concrete level to inspire improvements of concrete XML-based technologies.

Abstractly speaking, PA may be used as a way to describe the processes (at least their interfaces) to be composed. Afterwards, interactions are implicitly given by matching inputs and outputs of entities. In such a case, the PA parallel composition operators are used directly as a way to denote such compositions among entities: the global system is given as the parallel composition of the entities. Depending on the operator used and its underlying semantics, many communication models can be expressed (binary in CCS, n-ary in CSP, with data exchange in LOTOS, etc).

Let us illustrate with a classical example of communications between a requester and several providers written out in CCS. The requester sends requests, receives refusals and terminates when it receives an acceptance.

```
proc Requester =
      ( 'request.Requester     ***     emit a request and restart
      + refusal.Requester ) *** OR get  a refusal and restart
      + acceptance.nil        *** OR get an acceptance and terminate
```

On the other hand, a provider receives requests and replies a refusal or an acceptance.

```
proc Provider =                  *** recursively, receive a request and
                                 *** send either refusal or acceptance
         request. ( 'refusal.Provider  + 'acceptance.Provider )
```

Parallel compositions and restriction sets are used to describe possible systems made up of a requester and several providers (three below). For the composition question, they are the means to work out interactions among the involved entities. In this example, we do not specify explicitly receivers of actions (*e.g.* ProviderA.request). However, they are not essential at this level of abstraction, and the main goal is that the request be accepted (`acceptance` reception). A pictorial representation of these interacting services is represented in Figure 2.

```
*** we will impose synchronization on this set of action names:
set restSetC = {request,refusal,acceptance}

*** which restricts the composition of 1 requester and 3 providers:
proc SystemC = (Requester | Provider | Provider | Provider) \ restSetC
```

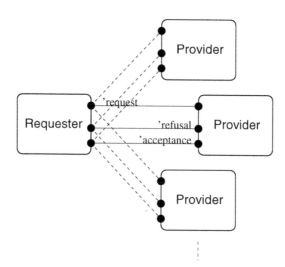

Fig. 2. An example of composition

In this subsection, we might imagine a description of service interfaces using another notation[4], the point of interest being the use of parallel composition semantics as the way to express interactions among processes.

3.4   Orchestration

Orchestration may be viewed as the problem of writing a central piece of code describing a possible way several web services may interact to work out a precise task. Such a more expressive and explicit notation is sometimes required to take more complex composition schemes into account. In such a case, PA may be used as an explicit orchestration language (BPEL is an executable XML-based language whose purpose is similar, but at the programming level). Interactions are expressed as one or several orchestrator processes and then the whole system is given as the parallel composition of the entities and the orchestrator(s).

    For illustration purposes, let us continue on our previous example. We imagine that we want for any reason to complete one transaction fired by a request

---

[4] The only assumption to be respected is that processes are described using a LTS-based model.

emission before trying another one. According to such a requirement and with respect to the existing services (Requester and Provider), we have to introduce an orchestrator whose goal is to control transactions and to reiterate the request until a provider answers positively.

```
proc Orchestrator =           *** Mutually recursive processes:
     request. (   'req1.Orch1   *** On request, send message to either
            + 'req2.Orch2 ) *** provider 1 or 2, and call the
                              *** corresponding process
proc Orch1 =    nok1.'refusal.Orchestrator
           + ok1.'acceptance.nil

proc Orch2 =    nok2.'refusal.Orchestrator
           + ok2.'acceptance.nil
```

In the example below, we introduce some processes interacting together. We rename actions (using the notation [newname/oldname, ...] meaning that all the occurrences of the action oldname in the behaviour are replaced by newname) as defined initially to distinguish the two transactions (with each provider instance). By the way, we stress that the use of such an orchestrator is helpful in case of heterogeneous definitions of services: it allows to compensate their differences (basic adaptations). We end in Figure 3 with an overview of these interacting entities.

```
set restSet0 = {request,refusal,acceptance,req1,ok1,nok1,req2,ok2,nok2}

*** we rename channels of the two providers in the composition:
proc System0 = (   Requester
              | Provider[req1/request,ok1/acceptance,nok1/refusal]
              | Provider[req2/request,ok2/acceptance,nok2/refusal]
              | Orchestrator) \ restSet0
```

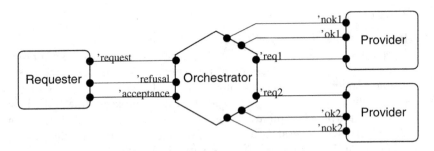

Fig. 3. An example of orchestration

In this subsection, we describe processes using PA, but it is not mandatory because PA is used here to express interactions among services. Basically, we would imagine to use WSDL interfaces as a simple description of services to be composed.

4   Related Works

In this paper, we advocate the use of PA at different levels and for different purposes (description, composition, reasoning, development). Our objective in this section is to compare existing works with the proposals at hand.

At this abstract level, many works originally tended to describe WSs and their composition using semi-formal notations, especially workflows [19]. More recently some more formal proposals grounded for most of them on transition system models (LTSs, Mealy automata, Petri nets) have been suggested [15, 27, 12, 2, 18]. PAs are adequate to describe WSs, because they allow to formally describe dynamic processes. Compared to the automata-based approaches, its main benefit is its expressiveness, particularly due to a large number of existing calculi enabling one to choose the most adequate formalism depending on its final use. Additionally, another interest of PA is that their constructs are adequate to specify composition due to their compositionality[5] property (not the case of the Petri net notation for instance). At last, textual notations (even if sometimes criticized to be less readable than transition systems) are more adequate to describe real-size problems, as well as to reason on them.

Regarding the reasoning issue, works have been dedicated to verifying WS descriptions to ensure some properties of systems [29, 10, 8, 27, 9, 26]. These works use model checking to verify some properties of cooperating WSs described using XML-based languages (DAML-S, WSFL, BPEL, WSCI). Accordingly, they abstract their representation and ensure some properties using ad-hoc or existing tools. As far as the composition is concerned, different techniques have emerged which ensure a correct composition such as automatic composition [2, 23], planning [22, 18] or model deduction [27]. However, most of the other proposals do not ensure this composition correctness [9, 15, 12]. The use of powerful proof theory accompanying PA helps to readily check equivalences between possible requests and composite WS, and then to ensure a correct composition. On a wider scale, PA are equipped with state-of-the-art reasoning tools, such as the SPIN model checker or the CADP toolbox, and hence are adequate to any kind of press-button reasoning.

Finally, we note that, to our knowledge, very few formal approaches have been proposed to develop certified WSs. The recent proposal of Lau and Mylopoulos [17] argues that TROPOS (an agent-oriented methodology) may be used to design WSs, but this paper does not take into account the formal part of the methodology [11]. Mantell [21] advocates a tool to map UML processes into BPEL ones, but the semi-formality of UML is a limit to the validation and verification stage. On the industrial side, platforms like .NET and J2EE make it possible to develop WSs, but they do not propose methods (above all formal) to achieve this goal.

---

[5] *Compositionality* corresponds to the possibility to build bigger behaviours from small ones.

5   Concluding Remarks

PA offers a wide panel of languages and tools to deal with many issues raised in WSs. While being based on a small core of usual operators, they propose many variants to deal with many specific aspects (data description, time, probability, mobility, etc) and involving different communication models.

When one uses a PA for reasoning purposes, it is worth noting that, while selecting the description language, an adequate trade-off should be chosen between expressiveness of the calculus *i.e.* the richness of WS features which can be described, and the verification abilities of the tool support accompanying such an algebra. In this direction, the goal is to abstract the system so as to reason on it using existing tools. Consequently, the more we abstract, the easier the reasoning is (but the less precise of course). A well-known example to justify this statement is the state explosion problem raised while trying to verify processes involving data descriptions.

A recent trend aims at using the π-calculus in the WS area. Regarding the directions mentioned herein, at first sight, the π-calculus does not seem adequate because: (i) the name passing is not of interest for the interface description, (ii) it is not possible to express the name passing into existing XML-based technologies (*e.g.* BPEL with regards to the development stage), (iii) it still lacks efficient and robust model checkers. However, these notions of name passing and dynamic instantiations are definitively of interest in the context of WSs (a new client arrives and has to tell the address where answers should be sent, the server forks and dynamically creates a sub-process parametrized with a new address, etc) and may be put forward in the next technology proposal step.

The results we have overviewed in this paper are at an early maturity stage and much work remains to be done before they are widely applicable to large-scale applications. Many perspectives can naturally be considered. A first one concerns the use of PA for WS development. Although it is promising due to the similar foundation of such abstract descriptions and some of the possible executable languages (such as BPEL), some work remains to be carried out to have at one's disposal a formal refinement allowing a correct encoding from one level to the other one. In this direction, the implementation of a prototype generating (skeletons of) executable code is considered as well. Another direction is to study a possible adjustment of existing equivalence checking algorithms so as to propose efficient automatic compositions of WSs. A related issue is to determine which compatibility checking (to ensure the possible composition of a set of WSs) can be performed using existing tools [3]. This raises the longer-term perspective of adaptation: when some services are not compatible, one can envision to automatically create adaptors, i.e. pieces of codes sitting between WSs and which ensure a correct communication. Finally, PA may be viewed as a great source of inspiration for the design of standard proposals in the context of WSs.

Acknowledgments

This work is partially supported by Project ASTRO funded by the Italian Ministry for Research under the FIRB framework (funds for basic research). The authors thank Antonella Chirichiello for her help on the part dedicated to the development of processes in BPEL.

References

1. T. Andrews, F. Curbera, H. Dholakia, Y. Goland, J. Klein, F. Leymann, K. Liu, D. Roller, D. Smith, S. Thatte, I. Trickovic, and S. Weerawarana. Specification: Business Process Execution Language for Web Services Version 1.1. 2003. Available at http://www-106.ibm.com/developerworks/webservices/library/ws-bpel/.
2. D. Berardi, D. Calvanese, G. De Giacomo, M. Lenzerini, and M. Mecella. Automatic Composition of E-services That Export Their Behavior. In M. E. Orlowska, S. Weerawarana, M. P. Papazoglou, and J. Yang, editors, *Proc. of ICSOC'03*, volume 2910 of *LNCS*, pages 43–58, Italy, 2003. Springer-Verlag.
3. L. Bordeaux, G. Salaün, D. Berardi, and M. Mecella. When are two Web Services Compatible? In *Proc. of TES'04*. To appear.
4. M. Bravetti and G. Zavattaro, editors. *Proc. of the 1st International Workshop on Web Services and Formal Methods (WS-FM'04)*, Italy, 2004. To appear in ENTCS.
5. A. Brogi, C. Canal, E. Pimentel, and A. Vallecillo. Formalizing Web Services Choreographies. In *Proc. of WS-FM'04*, Italy, 2004. To appear.
6. A. Chirichiello and G. Salaün. Developing Executable and Certified Web Services from Abstract Descriptions. Submitted.
7. E. M. Clarke, O. Grumberg, and D. Peled. *Model Checking*. The MIT Press, 2000.
8. A. Deutsch, L. Sui, and V. Vianu. Specification and Verification of Data-driven Web Services. In ACM, editor, *Proc. of PODS'04*, pages 71–82, Paris, 2004. ACM Press.
9. H. Foster, S. Uchitel, J. Magee, and J. Kramer. Model-based Verification of Web Service Compositions. In *Proc. of ASE'03*, pages 152–163, Canada, 2003. IEEE Computer Society Press.
10. X. Fu, T. Bultan, and J. Su. Analysis of Interacting BPEL Web Services. In *Proc. of WWW'04*, pages 621–630, USA, 2004. ACM Press.
11. A. Fuxman, L. Liu, M. Pistore, M. Roveri, and J. Mylopoulos. Specifying and Analyzing Early Requirements: Some Experimental Results. In *Proc. of RE'03*, USA, 2003. IEEE Computer Society Press.
12. R. Hamadi and B. Benatallah. A Petri Net-based Model for Web Service Composition. In K.-D. Schewe and X. Zhou, editors, *Proc. of ADC'03*, volume 17 of *CRPIT*, pages 191–200, Australia, 2003. Australian Computer Society.
13. J. Harrison. Verification: Industrial Applications. *Lecture at 2003 Marktoberdorf Summer School*, USA.
14. G. J. Holzmann. *The Spin Model Checker, Primer and Reference Manual*. Addison-Wesley, Reading, Massachusetts, 2003.
15. R. Hull, M. Benedikt, V. Christophides, and J. Su. E-Services: a Look Behind the Curtain. In ACM, editor, *Proc. of PODS'03*, pages 1–14, USA, 2003. ACM Press.
16. ISO. LOTOS: a Formal Description Technique based on the Temporal Ordering of Observational Behaviour. Technical Report 8807, International Standards Organisation, 1989.

17. D. Lau and J. Mylopoulos. Designing Web Services with Tropos. In *Proc. of ICWS'04*, pages 306–313, San Diego, USA, 2004. IEEE Computer Society Press.

18. A. Lazovik, M. Aiello, and M. P. Papazoglou. Planning and Monitoring the Execution of Web Service Requests. In M. E. Orlowska, S. Weerawarana, M. P. Papazoglou, and J. Yang, editors, *Proc. of ICSOC'03*, volume 2910 of *LNCS*, pages 335–350, Italy, 2003. Springer-Verlag.

19. F. Leymann. Managing Business Processes via Workflow Technology. *Tutorial at VLDB'01*, Italy, 2001.

20. Z. Manna and A. Pnueli. *Temporal Verification of Reactive Systems – Safety*. Springer, 1995.

21. K. Mantell. From UML to BPEL. IBM developerWorks report, 2003.

22. S. A. McIlraith and T. C. Son. Adapting Golog for Composition of Semantic Web Services. In D. Fensel, F. Giunchiglia, D. McGuinness, and M.-A. Williams, editors, *Proc. of KR'02*, pages 482–496, France, 2002. Morgan Kaufmann Publishers.

23. B. Medjahed, A. Bouguettaya, and A. K. Elmagarmid. Composing Web services on the Semantic Web. *The VLDB Journal*, 12(4):333–351, 2003.

24. G. Meredith and S. Bjorg. Contracts and Types. *Communications of the ACM*, 46(10):41–47, 2003.

25. R. Milner. *Communication and Concurrency*. International Series in Computer Science. Prentice Hall, 1989.

26. S. Nakajima. Model-checking Verification for Reliable Web Service. In *Proc. of OOWS'02, satellite event of OOPSLA'02*, USA, 2002.

27. S. Narayanan and S. McIlraith. Analysis and Simulation of Web Services. *Computer Networks*, 42(5):675–693, 2003.

28. J. Parrow. *An Introduction to the π-Calculus*, chapter 8, pages 479–543. Handbook of Process Algebra. Elsevier, 2001.

29. G. Salaün, L. Bordeaux, and M. Schaerf. Describing and Reasoning on Web Services using Process Algebra. In *Proc. of ICWS'04*, pages 43–51, San Diego, USA, 2004. IEEE Computer Society Press.

30. G. Salaün, A. Ferrara, and A. Chirichiello. Negotiation among Web Services using LOTOS/CADP. In M. Jeckle and L.-J. Zhang, editors, *Proc. of ECOWS'04*, volume 3250 of *LNCS*, Germany, 2004. Springer-Verlag. To appear.

31. W3C. *Web Services Choreography Description Language Version 1.0*. Available at http://www.w3.org/TR/2004/WD-ws-cdl-10-20040427/.

32. W3C. *Web Services Choreography Interface 1.0*. Available at http:www.w3.org/TR/wsci.

33. W3C. *Web Services Choreography Requirements 1.0 (draft)*. Available at http:www.w3.org/TR/ws-chor-reqs.

Flexible Coordination of E-Services

Roger S. Barga[1] and Jing Shan[2]

[1] Microsoft Research, Microsoft Corporation
Redmond, WA 98053
[2] College of Computer & Information Science,
Northeastern University,
Boston, MA 02115

Abstract. Information systems are increasingly being built using e-services invoked across the internet. Businesses can create a virtual application by composing simpler existing e-services provided by a number of different service providers. These composed e-services, or e-workflows, must operate in a highly fluid environment, where new services become available for use and business practices are subject to frequent change. This imposes demanding requirements for flexibility on the workflow system that manages the business process. In this paper we describe our efforts to extend workflow orchestration by incorporating information workers in process enactment, to enable both system and users to cooperate in interpreting the process model. This allows for a variety of methods for simple e-services to be interconnected and composed into more complex e-services. This approach also provides an elegant way to incorporate ad hoc changes or specialization during workflow enactment and to react to exceptions. We are building a prototype of our workflow adaptation services on top of a commercial workflow orchestration engine.

1 Introduction

Despite having ample reason to do so, very few businesses system-support the processes behind their bread and butter operations. Today business processes are typically run by information workers guided by process 'maps', often memorized, where the system offers no active support or enforcement. While workflow management systems have been around for years, their penetration into core processes of enterprise businesses is poor – only a few percent of candidate systems. One main reason is that when businesses have tried to build workflow systems that embody a process, in particular where involvement from information workers is required, the exercise has been painful and delivered poor return on investment. This failure of workflow management has partly been attributed to a lack of flexibility. The challenge is to support process automation and yet provide information workers the opportunity to add unique value to the process. This is essential for composition of e-services, where often the next step in a process flow depends on information available only at runtime; typically, information workers make these critical decisions based on runtime context. Consequently, flexible workflow management is now an active research topic.

M.-C. Shan et al. (Eds.): TES 2004, LNCS 3324, pp. 69–79, 2005.

Previous work in the area has examined how a workflow system can be extended and enhanced, how static workflows can be made *adaptive*. Research challenges for adaptive workflow include:

- Controlled handling of exceptions during execution; [Casati98, HG98]
- Support for process model change, migrating active instances from an old workflow schema to a newer version; [EKR95, RD98, HSB98, HSW01]
- Local adaptation or specialization of a workflow instance; [GSB+00, CIJ+00, BDS+02, ZFC+02]

While previous work in this area recognizes that change is a way of life in executing a business process, a basic premise underlying this research is still that work is repetitive and can be relatively completely prescribed. And while an understanding seems to have emerged that changes to the flow of a business process requires process definition and process enactment to be entwined, most efforts on adaptive workflow are based on the premise that the workflow enactment engine is solely responsible for interpreting the model [Jorg01]. In other words, information workers are free to contribute by making alterations to the process model, but not by actively interpreting aspects of the model at runtime. Thus, the process model must be formally complete to prevent ambiguity and deadlock from paralyzing process execution. This view complicates process model design, because all variants must be included *a priori*. Completely describing the model and possible variants is always harder than first thought – exception cases start coming out of the woodwork, and then exceptions to the exception cases. Describing a model in advance also precludes typical customizations required to deploy a system, first by the ISV and then by end users. In summary, these model-centric approaches to supporting flexible workflows suffer where information workers are involved, or where there needs to be more than one way of thinking about a process.

We believe that flexible workflow management requires both automation and information worker interactivity. We are exploring an approach to workflow enactment in which system and users interact in interpreting the model and handing exceptions that may arise. While the system makes decisions about prescribed activities and enforces correctness guarantees, users are free to select from available options or compose activity code on the fly to resolve ambiguities and handle special cases. Subject to less strict rules the articulation of the workflow can be more fluent; hence, the gap between modeled and real business processes can be decreased. More importantly, when facing changes in business practices, unexpected events or exceptions, information workers will not abandon the system if they are able to change its runtime configuration.

1.1 Application Requirements for Flexible Workflow Processing

To better understand the requirements for flexible workflow support, our company spoke to a large number of their small business and enterprise customers that deploy and support workflow based applications and services. One of the more common recurring patterns in their business requirements is support for "human intervention".

First, let's consider the instance of a business process. A clear requirement is that information workers want the ability to *intervene in the process at any point.* These users see their ability to be flexible as a competitive advantage, and do not want software to get in the way of this ability. These information workers that interact with the process make key business decisions, resolve problems, clarify expectations, and handle changes and track progress at each step along the way. These users also expressed the need to *invoke ad hoc activities on demand* as the business process unfolds, possibly taking advantage of new e-service types or instances.

Second, we found that business processes themselves – not the design, not the instance, tend to have a relatively short life span. Information workers tend to change them over time, such as in continuous process improvement programs and process reengineering. Consequently, support for composed e-services must allow for rapid and continual change after initially being deployed.

The requirements we have for flexible e-service coordination are:

1. We want the system to include an understanding of business goals, and be able to guide users towards achieving them;
2. We want to maximize the freedom experienced by users, while at the same time ensuring that they don't violate the business constraints. If a new type of e-service is available then an information worker should be free to include it, as long as it does not violate correctness constraints;
3. We want to model the process in a way that keeps the map between the business problem and model as simple as possible, in order to maximize fidelity to the business requirement.
4. We want to vary the level of detail we put into the model, so as to make sure we maximize the return on investment in our initial analysis and deployment efforts;
5. We want a modeling technique that supports customization, so that product builders, ISVs and customers are all enabled to add their requirements to a system in a way consistent with their capabilities;

In short, we need a workflow system that supports flexible and correct human interaction with the system at runtime.

2   Interactive Enactment for Flexible Workflows

The core components of our baseline workflow management system are illustrated in Figure 1. In this section we highlight our observations, insights and approach to extend these components to support interactive enactment of a business processes.

Examples of the type of interaction we seek to support include the following:

- Detect and resolve **business exceptions** generated by errors in message contents, message sequencing/omission problems or exceptions to business policy;
- Manually **moving the process** forward or backward one or more steps;
- **Canceling a process** at any time;

+ **Injecting process steps** on the fly to handle additional process steps deemed necessary by the information worker (e.g., case escalations);
+ Including **unstructured correspondence** within the process;
+ Allowing process steps to be fulfilled by **manual actions** or **unstructured correspondence**;

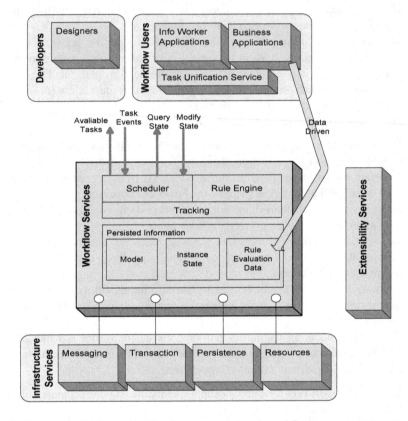

Fig. 1. Illustration of primary functional components in our workflow management system

These requirements for interactive enactment impose some significant design challenges to extend the components of a workflow system. One of the more significant challenges is to enhance the model for process definitions that supports both automation and interactivity. Support for unstructured correspondence, for example, presents challenges in the area of correlation and storage. The design of the model definition is also a challenge in terms of the breadth of meta-data that must be included to open aspects of the system for interaction. In addition, we must provide the ability for information workers to dynamically alter an active instance of a workflow at runtime, to react to changing business conditions or exceptions. Lastly, the design of visual tools to track and interact with an active workflow present their own design challenges.

2.1 Activity Representation

The workflow engine on which we are basing our implementation activates a process model to coordinate the execution of units of work called activities. In most systems, enactment is completely specified by the process model. Such strict adherence to a prescribed model makes it difficult to adapt or respond to unforeseen circumstances. In contrast, interactive enactment allows intertwined articulation and activation of an evolving online model. Change and exception handling is accommodated either by user-controlled activation (re-interpretation) or by updating the model at runtime. Because user interaction is allowed in activation and not just in modeling, the process model need not be 100% complete – low fidelity models are permitted. The low-fidelity model, which utilizes placeholder activities [GSB00] and interaction points, captures the essence of the process while abstracting details that can not be fully specified at design time. Placeholder activities identify the position of an activity in the flow, along with input and output signatures and optional correctness constraints. An interaction point is a breakpoint in process execution where information workers can inject new activities or adjust work items. The situated nature of the interaction further enhances the usability of such process models. At the extreme, we allow a models with no structure, just an unordered list of activities.

Activities are the scheduling primitives for our runtime – these are the actions done by one person at one place and at one time, such as sending an order request to an e-service or recording shipment of an order. Each activity provides well-understood behavior and explicit state reliance and impact. The system is modeled as a set of nodes, each representing an activity or composite activity in the context of a business goal. Each node includes the following attributes (*other details not related to the subject of this paper are omitted for brevity*):

- An *'available'* expression that specifies process and resource events that can trigger the activity (i.e., when the activity can be applied);
- A *guard* predicate that specifies runtime state that must satisfied before activity execution can proceed;
- An *'optional'* expression, to indicate if an activity is mandatory or optional;
- An optional *'provides'* expression (resource);
- An optional *'requires'* expression (resource);
- The **body**, which is the executable code block for the activity;
- An optional *'terminate'* expression that identifies conditions under which the activity should be terminated;
- A *'status'* variable indicating if the activity is complete, terminated, deferred, in process, etc.;
- A *correctness constraint* that identifies one or more named predicate(s) that must be satisfied after activity execution;
- A list of conditional *triggers* to be fired upon completion of the activity;

Enactment is driven by events, classified as either *process* events or *resource* events. Process events signal action initiation and completion, and are generated directly by activities as a consequence of performing activities. Resource events reflect changes in the environment, such as creation, deletion, and modification of resources, and temporal events such as deadlines or alarms. A coordinating process is responsible for evaluating activity expressions against the event stream – the result is a list of activities that must be performed to reach a given business goal, and a list of activities currently valid to perform in the context of that goal. Guards are expressions consisting of a reference to named rules. The rules and guards are part of the standard run-time package, which an ISV can extend or modify over time.

The completion of an activity can make a number of additional events available for an activity to perform: a process event signals the completion of an activity using triggers, which makes the next activity in the nominal control flow available. In addition, the completion of an action may include the creation or modification of one or more resources as side-effects, generating resource events making additional actions available. The coordinating process will then re-evaluate the expressions and produce a new list of activities, possibly recommending one for the benefit of novice users, and also show other actions available as the result of resource events. If there are multiple possible activities then the *"available list"* is presented to the user in an action pane or other form. In addition to providing this list of activities, the coordinator permits a user with proper privileges to manually supply an activity to respond to the event; a user interface is available for the information worker to compose an activity at runtime and correctness is enforced by constraints that specify guarantees that must be upheld after it is executed. This allows qualified users to intervene in the process at any point, to select newly available e-services, invoke ad hoc processes on demand, or react to exceptions that arise during normal processing.

The constraints on activity transitions are expressed using *requires* and *provides* statements. These are predicates that express inputs and outputs of each activity in the process, and thus the data pre-conditions and post-conditions that exist at each step. If at runtime a user wishes to skip a standard step on which a later step depends, a built-in guard will detect the data dependency. We provide additional nodes, representing manual data input processes, whose 'available' expression is the inverse of the guard clauses. Then, when progress is blocked a UI automatically becomes available that prompts the user to supply the missing data so the process can proceed.

2.2 Exception Handling

In general, most business processes consist of repeatable patterns that can be modeled in advance. However, while the expected path is generally quite simple, it is the number of exception paths as well as the number of operations such as change and cancel that can occur at any time during the life-cycle that creates considerable complexity when one tries to flow-chart the process. We refer to the expected path as the "happy path". Business exceptions cause the process to take a detour off the happy path. Examples of business exceptions include the following:

- Confirmations not being received on time – may be resolved by continuing to wait or moving the workflow forward, skipping the confirmation step;
- Errors in shipping codes on a sales order – may be resolved by correcting the codes, accepting them as is or rejecting the order;

⧫ A purchase order amount that exceeds business policy – may be resolved by changing the order or initiating an approval process;

⧫ An order cannot be fully confirmed – may be resolved by accepting the partial order or renegotiating the order with the vendor;

Exceptions are resolved by information workers that track the order and guide it thru its life cycle. Business decisions that affect the process are made by these information workers on a continual basis as they manage each process. We discovered that the processes in use by these customers must allow for information worker intervention at any stage. During these interactions, the process can be diverted down a different path, pushed ahead to a different state or held back. We also discovered that unstructured correspondence, such as emails and faxes, plays a vital role in these processes. Unstructured correspondence is needed to supplement, or even replace, actions in a modeled process. We address this issue again later in this section.

2.3 Supporting Adaptation of Active Workflow Instances

As previously described, exceptions and user interaction are possible with every step in a business process. The information worker must be free to skip or repeat actions. Changes and cancellations are allowed at any stage of the process. Validation and business policy exceptions requiring human intervention can occur with every message sent or received. These requirements quickly turn a process with only a handful of actions into a complex set of possible flows. One way we deal with this is to allow the information worker the ability to adapt a running workflow instance at runtime.

To better understand the requirements for workflow adaptation, let's consider the possible states of a running workflow illustrated in Figure 2. The running instance of a workflow model contains a description of activities that have *already been completed*, that are active and *currently being worked on,* and that are *planned but not yet carried out.* It is desirable that (1) already planned parts of a workflow model can be easily adapted to a new situation (the system supports dynamic process flow "replanning") and (2) parts of the process can be "undone" and enactment can restart at an earlier point (the system supports task "redo"). In addition, there is (3) the need for additional activities not specified in the workflow model but nevertheless must take place (the workflow system supports the insertion of "ad-hoc" activities).

Providing support for *replanning, redo* and *ad-hoc* activities constitutes our baseline support for providing flexibility by dynamic adaptation of running workflow instances. For workflow adaptation, we distinguish five ways (operators) in which the routing of cases along tasks can be changed:

Extend – Adding new tasks to the flow which (1) can be executed in parallel, (2) offer new alternatives (tasks), or (3) are executed in-between existing tasks;

Replace – A task is replaced by another task or an activity (i.e., refinement), or a complete region (activity flow) is replaced by another region (activity flow);

Reorder – Changing the order in which tasks are executed without adding new tasks, e.g., swapping tasks or making a process more or less parallel;

Remove – Removal of tasks or the removal of a complete region from the work-flow. Note, that task execution can be deferred using the reorder feature;

Redo – One or more completed activities in the instance are "undone" and enact-ment is restarted from an earlier point in the workflow model.

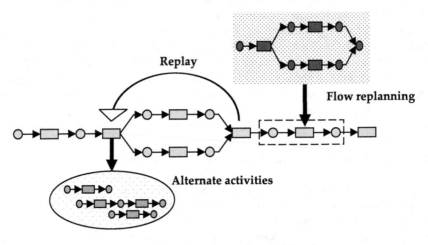

Fig. 2. Runtime alteration of an active workflow instance

2.4   User Visibility into Every Action

Users track and interact with business processes at a detailed level. They may review and scrutinize every message sent and received – they need to be aware of every state transition. The most basic requirement here for interactive enactment is that the state of a business process may be tracked by the system and be queriable. However, there is a little more to it than that. To support interactivity, the results returned when one queries the state of a process must make sense to the end user. The results must be in terms of actions and process state that the user will recognize. This allows develop-ment of a user interface that is driven by the process definitions themselves. The cur-rent process state and expected user actions can be rendered by reflecting on the proc-ess definitions and current process state. This is a requirement if one hopes to be able to support many different process definitions with a single visual tool.   So it seems natural to define processes using a single model that drives both process execution and end-user visualization.

2.5   Correlation and Storage

Actions and communications that take place within the scope of a business process need to be correlated into a common *process context*.   Correlation needs to handle both structured messages and unstructured correspondence.   Structured messages are generally business documents that are exchanged between partners – for example, a purchase order. Structured messages could also be used for intra-company communication – for example, an approval form. Structured messages also have embedded context information that can be used for automatic correlation into the correct process context – for example, "processed" or "agreementID" fields. In contrast, unstructured correspondence is generated by information workers as free-form emails, faxes and notes. Unstructured correspondence is generally not computer readable and lacks the embedded context information needed for automatic correlation. Despite these limitations, unstructured correspondence plays a vital role in interactive business process enactment and must be supported within those processes.

2.6 Implementation Approach

To address the challenges of supporting the functional requirements of interactive workflow enactment, we are pursuing a rule-based approach for dynamic adaptation and runtime verification of correctness constraints (business rules). Our current system design includes distinct functional modules that extend the underlying workflow engine. First, a monitoring module listens to workflow data and control events raised by the workflow engine and determines which events represent adaptation or exception events. Second, there is a control module that is responsible for determining which workflow instances are affected by the event. Third, there is an adaptation module responsible for modifying the affected workflow instances, possibly removing or inserting activities, or presenting alternate activities for exception handling, so the workflow instance can better cope with the new situation. Finally, a monitoring module is responsible for checking whether the assumptions of the adaptation module are matched when the adapted workflow is executed (business rules and process correctness rule verification). We chose this modular architecture to provide a separation of concerns, dividing distinct logical functionality into different modules, thus allowing the replacement and further refinement of individual modules as we apply our approach to different application domains.

3 Refining and Extending the Model

In summary, the more innovative aspects of our proposed extensions for workflow enactment are: i) support for low fidelity models, ii) interactive process enactment, which allows information workers to participate in interpreting the model at runtime and possibly define activities on the fly with rich UI support, and iii) automatic enforcement of correctness guarantees to ensure that the workflow generated at runtime matches the business constraints.

There are many details behind our approach and underlying runtime that we could not cover in this paper. One area we are actively pursuing is the definition and enforcement of correctness guarantees. Currently we enforce data dependencies between activities and the execution of activities through the use of guards and constraints. In addition, we enforce correctness constraints (named predicates) on activity execution at runtime, but are still exploring alternatives for the specification and runtime enforcement of correctness for the complete business process (flow).

4   A Comparison with Related Work

Dynamic change in workflow management systems, as well as correctness issues related to it, was first proposed in [EKR95]. Since then, the importance of flexibility in workflow systems has been widely recognized [Casati98, BDK99, RD98, HSB98, WASA, NUTT96, GSB+00, HSW01]. A classification of pas work in the area of adaptive workflow is given in [HSB98]. Recently, with the rapid development of web services, the dynamic character of these systems demands even more flexibility from its underlying workflow engine [CIJ+00, BDS+02].

In general, there are two approaches towards dynamic workflow adaptation. The first approach is to predefine the entire process. Schema changes are realized through either a set of primitives [RD98, Casati98, HK96] or late-binding/open-points technique [GSB+00]. The second approach is to apply soft constraints, typically defined as rules, to guide the workflow [ZFC+02, OYP03]. The latter approach does not require the workflow designer to figure out every detail in advance; hence, it allows more flexibility. Our model takes the latter approach to provide dynamic workflow adaptation.

WIDE [Casati98] and ADEPT [RD98] both proposed a set of primitives to realize schema evolutions. ECA rules are used in WIDE to handle expected exceptions. OPERA [HG98, HG00] applies programming language concepts to workflow systems with a focus on the execution environment. All these approaches assume the workflow schema is predefined. As a contrast, in our approach there is no complete specification of the entire workflow at design time. The flow is defined gradually at run time by allowing user-specified rules.

In CMI [GSB00], process templates and placeholders are used during design time to allow further specification and extension at run time. However, the input and output signature of the placeholder is fixed, which limits flexibility. In contrast, our approach allows new rules to be added at any point and the resulting structure changes are transparent to users.

E-service compositions are modeled as processes in workflow systems. The dynamic nature of web services requires more flexibility than traditional workflows. In eFlow [CIJ +00], dynamic adaptation is realized through generic nodes and service selection rules. Their adaptation focuses more on the dynamic selection of service providers rather than the dynamic change in process schema as in our model. In [BDS+02], service compositions are represented in a state chart. Each process is executed in a decentralized environment. A rule engine is used in [OYP03] to govern and guide service composition processes. However, no detail implementation of schema evolution is given.

References

[CIJ+00]   Adaptive and Dynamic Service Composition in eFlow by Fabio Casati, Ski Il-nicki, Lijie Jin, Vasudev Krishnamoorthy, Ming-Chien Shan, Hewlett-Packard Tech Report 2000.

[Casati98]   Models, Semantics and Formal Methods for the design of workflows and their exceptions by Fabio Casati, PhD thesis, 1998.

[EKR95]   Dynamic Change within Workflow Systems by Clarence Ellis, Karim Keddara and Grzegorz Rozenberg COOCS 95.

[EL95]   Workflow Activity Model WAMO, Johann Eder and Walter Liebhart CoopIS 95.

[OYP03]   A Framework for Business Rule Driver Service Composition, by Bart Orriens, Jian Yang, and Mike P. Papazoglou, TES-03.

[RD98]   ADEPTflex – Supporting Dynamic Changes of Workflows Without Loosing Control, by Manfred Reichert and Peter Dadam, Journal of Intelligent Information Systems March/April 1998.

[JH99]   Managing Evolving Workflow Specifications with Schema Versioning and Migration Rules, Gregor Joeris and Otthein Herzog, TZI Technical Report 1999.

[Jorg01]   Interaction as a Framework for Flexible Workflow. Havard D. Jorgensen, Proceedings of 2001 International ACM SIGGROUP Conference on Supporting Group Work, 2001.

[HG98]   Flexible Exception Handling in the OPERA Process Support System by Claus Hagen and Gustavo Alonso, ICDCS'98.

[HG00]   Exception Handling in Workflow Management Systems, by Claus Hagen and Gustavo Alonso, IEEE Transactions on Software Engineering Oct 2000.

[HSB98]   A Taxonomy of Adaptive Workflow Management, by Yanbo Han, Amit Sheth and Christopher Bussler, CSCW98.

[GSB+00]   Managing Escalation of Collaboration Processes in Crisis Mitigation Situations Dimitrios Georgakopoulos, Hans Schuster, Donald Baker and Andrzej Cichocki, ICDE'00.

[BDK99]   Towards Adaptive Workflow Systems, CSCW-98 Workshop Report. By Abraham Bernstein, Chrysanthos Dellarocas and Mark Klein. SIGMOD Record 28,1999.

[WASA]   Weske, M.: Flexible Modeling and Execution of Workflow Activities. Technical Report Angewandte Mathematik und Informatik 08/97-I, University of Muenster, 1997. (http://dbms.uni-muenster.de/menu.php3?item=projects&page='wasa/index.php3?id=1').

[NUTT96]   The evolution toward flexible workflow systems (1996) by Gary J. Nutt.

[HK96]   ObjectFlow: Towards a process management infrastructure, by Hsu and Kleissner, Distributed and Parallel Databases 1996.

[HSW01]   Flexible Workflow Management in the OPENflow System, by J.J. Halliday, S. K. Shrivastava and S. M. Wheater, IEEE International Enterprise Distributed Object Computing Conference, EDOC'01.

[BDS+02]   Declarative Composition and Peer-to-Peer Provisioning of Dynamic Web Services, Boualem Benatallah, Marlon Dumas, Quan Z. Sheng and Anne H.H.Ngu, ICDE'02.

[ZFC+02]   PLMflow – Dynamic Business Process Composition and Execution by Rule Inference, by Liangzhao Zeng, David Flaxer, Henry Change and Jun-Jang Jeng. TES 2002.

$\mathcal{ESC}$: A Tool for Automatic Composition of e-Services Based on Logics of Programs

Daniela Berardi[1], Diego Calvanese[2], Giuseppe De Giacomo[1],
Maurizio Lenzerini[1], and Massimo Mecella[1]

[1] Università di Roma "La Sapienza",
Dipartimento di Informatica e Sistemistica "Antonio Ruberti",
Via Salaria 113, 00198 Roma, Italy
{berardi, degiacomo, lenzerini, mecella}@dis.uniroma1.it
[2] Libera Università di Bolzano/Bozen,
Facoltà di Scienze e Tecnologie Informatiche,
Piazza Domenicani 3, 39100 Bolzano/Bozen, Italy
calvanese@inf.unibz.it

Abstract. In this paper we discuss an effective technique for automatic service composition and we present the prototype software that implements it. In particular, we characterize the behavior of a service in terms of a finite state machine. In this setting we discuss a technique based on satisfiability in a variant of Propositional Dynamic Logic that solves the automatic composition problem. Specifically, given *(i)* a client specification of his *desired service*, i.e., the service he would like to interact with, and *(ii)* a set of available services, our technique synthesizes the orchestration schema of a composite service that uses only the available services and fully realizes the client specification. The developed system is an open-source software tool, called $\mathcal{ESC}$ (E-Service Composer), that implements our composition technique starting from services, each of them described in terms of a WSDL specification and a behavioral description expressed in any language that can capture finite state machines.

1 Introduction

One of the basic aspects of the *Service Oriented Computing*, and of the Extended Service Oriented Architecture proposed by [19], is the composition of services. Basically, service composition addresses the situation when a client request cannot be satisfied by any (single) available service, but a *composite* service, obtained by combining "parts of" available *component* services, might be used [18, 11, 7].

Service composition involves two different issues. The first, referred to as *composition synthesis* is concerned with synthesizing such a new composite service, thus producing a specification of how to coordinate the component services to obtain the composite service. Such a specification can be obtained either *automatically*, i.e., using a tool that implements a composition algorithm, or

M.-C. Shan et al. (Eds.): TES 2004, LNCS 3324, pp. 80–94, 2005.

manually by a human, possibly with the help of CASE-like tools. In what follows, we will refer to such a specification of the composite service as *orchestration schema*, according to [1]. The second issue, referred to as *orchestration*, is concerned with coordinating, during the composite service execution, the various component services according to the orchestration schema previously synthesized, and also monitoring control and data flow among the involved services, in order to guarantee the correct execution of the composite service. Such activities are performed by the *orchestration engine* [1].

It has been argued [19, 1], that in order to be able to automatically synthesize a composite service starting from available ones, the available services should provide rich service descriptions, consisting of *(i)* interface, *(ii)* capabilities, *(iii)* behavior, and *(iv)* quality. In particular, the service interface description publishes the service signature[3], while the service capability description states the conceptual purpose and expected results of the service. The (expected) behavior of a service during its execution is described by its service behavior description. Finally, the Quality of Service (QoS) description publishes important functional and non-functional service quality attributes[4].

Several works in the service literature (refer to [17] for a survey) address the problem of service composition in a framework where services are represented in terms of their (static) interface. The aim of this work is twofold: first, we discuss an effective technique for automatic service composition, when services are characterized in terms of their behavior, and then we present the prototype design and development of an open source software tool implementing our composition technique, namely $\mathcal{ESC}$ (E-Service Composer)[5].

In [8, 7] we have devised a framework where services export their behavior as finite state machines, and in [7] we have developed an algorithm that, given *(i)* a client specification of his *desired service*, i.e., the service he would like to interact with, and *(ii)* a set of available services, synthesizes the orchestration schema of a composite service that uses only the available services and fully realizes the client specification. We have also studied the computational complexity of our algorithm: it runs in exponential time with respect to the size of the input state machines. Observe that, it is easy to come up with examples in which the orchestration schema is exponential in the size of the component services. However, practical experimentation conducted over some real cases with the prototype, shows that, given the complexity of the behavior of real services, the tool can effectively build a composite service.

Although some papers have already been published that discuss either behavioral models of services ([17]), or propose algorithms for computing composition

---

[3] E.g., as a WSDL file.

[4] E.g., service metering and cost, performance metrics (e.g., response time), security attributes, (transactional) integrity, reliability, scalability, availability, etc.

[5] cf. the PARIDE (Process-based frAmewoRk for composItion and orchestration of Dynamic E-services) Open Source Project: http://sourceforge.net/projects/paride/ that is the general framework in which we intend to release the various prototypes produced by our research.

(e.g., [18, 11, 20]), to the best of our knowledge, our research is the first one tackling *simultaneously* the following issues: *(i)* presenting a formal framework where the problem of service composition is precisely characterized, *(ii)* providing techniques for automatically computing service composition in the case of services represented as finite state machines and, *(iii)* implementing our composition technique into an effective software tool.

The rest of the paper is organized as follows. In Section 2 we discuss our framework for services that export their behavior. In Section 3 we present our technique for automatic service composition. In Section 4 we describe our tool. Finally, in Section 5 we draw conclusions and discuss future work.

2   General Framework

A service is a software artifact that interacts with its client and possibly other services in order to perform a specified task. A client can be either a human or a software application. When executed, a service performs a given task by executing certain actions *in coordination* with the client.

We characterize the exported behavior of a service by means of an *execution tree*. The nodes of such a tree represent the sequence of actions that have been performed so far by the service, while the successor nodes represent the actions that can be performed next at the current point of the computation. Observe that in such an execution tree, for each node we can have at most one successor node for each action. The root represents the initial state of the computation performed by the service, when no action have been executed yet. We label the nodes that correspond to completed execution of the service as "final", with the intended meaning that in these nodes the service can (legally) terminate.

We concentrate on services whose behavior can be represented using a *finite number of states*. We do not consider any specific representation formalism for representing such states (such as action languages, situation calculus, statecharts, etc.). Instead, we use directly deterministic finite state machines (i.e., deterministic and finite labeled transition systems). FSMs can capture an interesting class of services, that are able to carry on rather complex interactions with their clients, performing useful tasks. Moreover, several papers in the service literature adopt FSMs as the basic model of exported behavior of services [17, 1]. Also, FSMs constitute the core of statecharts, which are one of the main components of UML and are becoming a widely used formalism for specifying the dynamic behavior of entities.

The alphabet of the FSM (i.e., of the symbol labeling transitions) is formed by the actions that the service can execute. Such actions are the abstractions of the effective input/output messages and operations offered by the service. As an example, consider a service that allows for searching and listening to mp3 files; in particular, the client may choose to search for a song by specifying either its author(s) or its title (action `search_by_author` and `search_by_title`, respectively). Then the client selects and listens to a song (action `listen`). Finally, the client chooses whether to perform those actions again. The WSDL interface of this service and the finite state machine describing its behavior are reported in Figure 1[6].

To represent the set of services available to a client, we introduce the notion of *community* $\mathcal{C}$ of services, which is a (finite) set of services that share a common (finite) set of actions Σ, also called the *alphabet* of the community. Hence, to join a community, a service needs to export its behavior in terms of the alphabet of the community. From a more practical point of view, a community can be seen as the set of all services whose descriptions are stored in a repository. We assume that all such service descriptions have been produced on the basis of a common and agreed upon reference alphabet/semantics. This is not a restrictive hypothesis, as many scenarios of cooperative information systems, e.g., e-Government [4] or e-Business [12] ones, consider preliminary agreements on underlying ontologies, yet yielding a high degree of dynamism and flexibility.

Given a service A_i, the execution tree $T(A_i)$ *generated* by A_i is the execution tree containing one node for each sequence of actions obtained by following (in any possible way) the transitions of A_i, and annotating as final those nodes corresponding to the traversal of final states.

When a client requests a certain service from a service community, there may be no service in the community that can deliver it directly. However, it may be possible to suitably orchestrate (i.e., coordinate the execution of) the services of the community so as to provide the client with his desired service. In other words, there may be an orchestration that coordinates the services in the community, and that realizes the client desired service.

Formally, let the community $\mathcal{C}$ be formed by n services $A_1, \ldots, A_n$. An orchestration schema O of the services in $\mathcal{C}$ can be formalized as an *orchestration tree* $T(O)$:

- The root ε of the tree represents the fact that no action has been executed yet.
- Each node x in the orchestration tree $T(O)$ represents the history up to now, i.e., the sequence of actions as orchestrated so far.
- For every action a belonging to the alphabet Σ of the community and $I \in [1..n]$ [7] $(1, \ldots, n$ stand for the services $A_1, \ldots, A_n$, respectively), $T(O)$ contains at most one[8] successor node $x \cdot (a, I)$.

---

[6] Final nodes are represented by two concentric circles.

[7] We use $[i..j]$ to denote the set $\{i, \ldots, j\}$.

[8] Note that in our framework we focus on *actions* that a service may execute. Therefore, at this level of abstraction each action has a well-determined functionality. Observe also that we have avoided introducing data at the level of abstraction presented in this paper: in this way the complexity which is intrinsic in the data does not have a disruptive impact on the complexity which is intrinsic in the process. In fact, introducing data in a naive way is possible in our setting (e.g., by encoding data within the state) but it would make composition exponential in the data. This is considered unacceptable: the size of data is typically huge (wrt the size of the services) and therefore the composition should be kept polynomial in the data. In the future we will study how to add data to our framework by taking such observations into account.

```
<definitions ...
   xmlns:y="http://new.thiswebservice.namespace"
   targetNamespace="http://new.thiswebservice.namespace">

   <!-- Types -->
   <types>
      <element name="ListOfSong_Type">
         <complexType>
            <sequence>
               <element minOccurs="1"
                        maxOccurs="unbound"
                        name="SongTitle"
                        type="xs:string"/>
            </sequence>
         </complexType>
      </element>
   </types>

   <!-- Messages -->
   <message name="search_by_title_request">
      <part name="containedInTitle" type="xs:string"/>
   </message>
   <message name="search_by_title_response">
      <part name="matchingSongs" xsi:type="ListOfSong_Type"/>
   </message>
   <message name="search_by_author_request">
      <part name="authorName" type="xs:string"/>
   </message>
   <message name="search_by_author_response">
      <part name="matchingSongs" xsi:type="ListOfSong_Type"/>
   </message>
   <message name="listen_request">
      <part name="selectedSong" type="xs:string"/>
   </message>
   <message name="listen_response">
      <part name="MP3fileURL" type="xs:string"/>
   </message>

   <!-- Service and Operations -->
   <portType name="MP3ServiceType">
      <operation name="search_by_title">
         <input message="y:search_by_title_request"/>
         <output message="y:search_by_title_response"/>
      </operation>
      <operation name="search_by_author">
         <input message="y:search_by_author_request"/>
         <output message="y:search_by_author_response"/>
      </operation>
      <operation name="listen">
         <input message="y:listen_request"/>
         <output message="y:listen_response"/>
      </operation>
   </portType>

</definitions>
```

(a) WSDL

$a = search\_by\_author$
$t = search\_by\_title$
$l = listen$

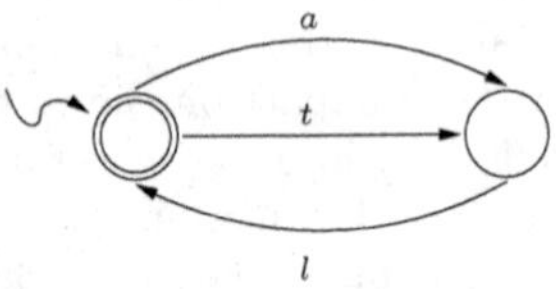

(b) FSM

Fig. 1. The MP3 service

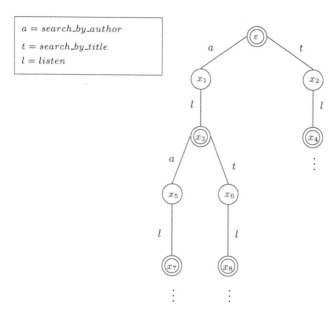

Fig. 2. Client specification as a tree

- Some nodes of the orchestration tree are annotated as *final*: when a node is final, and only then, the orchestration can be legally stopped.
- We call a pair $(x, x \cdot (a, I))$ an *edge* of the tree. Each edge $(x, x \cdot (a, I))$ of $T(O)$ is labeled by a pair (a, I), where a is the orchestrated action, $I \in [1..n]$ denotes the nonempty set of services in $\mathcal{C}$ that execute the action.
  As an example, the label $(a, \{1, 3\})$ means that the action a requested by the client is executed by, more precisely delegated to, the services A_1 and A_3.

Given an orchestration tree $T(O)$ and a path p in $T(O)$ starting from the root, we call the *projection* of p on a service A_i the path obtained from p by removing each edge whose label (a, I) is such that $i \notin \{I\}$, and collapsing start and end node of each removed edge.

We say that an orchestration O is *coherent* with a community $\mathcal{C}$ if for each path p in $T(O)$ from the root to a node x and for each service A_i of $\mathcal{C}$, the projection of p on A_i is a path in the execution tree $T(A_i)$ from the root to some node y, and moreover, if x is final in $T(O)$, then y is final in $T(A_i)$.

In our framework, we define *client specification* a specification of the orchestration tree according to the client desired service. Of the orchestration tree, the client *only* specifies the actions he would like to be executed by the desired service. Figure 2 shows a (portion of an infinite) orchestration tree representing the client specification: note that the edges of the tree are labeled only by actions. The client specification can be *realized* by an orchestration tree only if it is possible to find a suitable labeling for each action with a non empty set I of (identifiers of) services that can execute it. In this

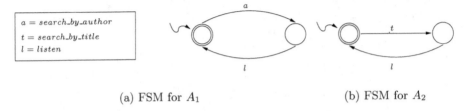

$a = search\_by\_author$

$t = search\_by\_title$

$l = listen$

(a) FSM for A_1 (b) FSM for A_2

Fig. 3. Services in the community

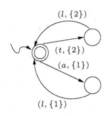

Fig. 4. Composition of A_0 wrt A_1 and A_2

work, we consider specifications that can be expressed using a finite number of states, i.e., as FSMs.

Given a community C of services, and a client specification A_0, the problem of *composition existence* is the problem of checking whether there exists an orchestration schema that is coherent with C and that realizes A_0. The problem of *composition synthesis* is the problem of synthesizing an orchestration schema that is coherent with C and that realizes A_0.

Since we are considering services that have a finite number of states, we would like also to have an orchestration schema that can be represented with a finite number of states, i.e., as a Mealy FSM (MFSM), in which the output alphabet is used to denote which services execute which action.

As an example, consider the case in which the service community is constituted by two services, A_1 and A_2, whose behaviors/FSMs are shown in Figure 3. A_1 allows for searching for a song by specifying its author(s) and for listening to the song selected by the client; then, it allows for executing these actions again. A_2 behaves like A_1, but it allows for retrieving a song by specifying its title.

If the client specification is the FSM shown in Figure 1(b)[9], then a composition exists, and its orchestration schema is the Mealy FSM shown in Figure 4, in which all the actions requested by the client are delegated to services of the community. In particular, the execution of search_by_author action and its subsequent listen action are delegated to A_1, and the execution of search_by_title action and its subsequent listen action to A_2.

---

[9] Note that it compactly represents the tree in Figure 2.

3   Automatic Service Composition

In the framework presented in the previous section, we are interested in knowing whether: (i) it is always possible to check the existence of a composition; (ii) if a composition exists, there exists an orchestration schema which is a finite state machine, i.e., a *finite state composition*; (iii) if a finite state composition exists, how to compute it. Our approach is based on reformulating the problem of service composition in terms of satisfiability of a suitable formula of Deterministic Propositional Dynamic Logic (DPDL [15]), a well-known logic of programs developed to verify properties of program schemas. DPDL enjoys three properties of particular interest: (i) the *tree model property*, which says that every model of a formula can be unwound to a (possibly infinite) tree-shaped model; (ii) the *small model property*, which says that every satisfiable formula admits a finite model whose size is at most exponential in the size of the formula itself; (iii) the EXPTIME-completeness of satisfiability in DPDL.

We represent the FSMs of both the client specification A_0 and the services $A_1, \ldots, A_n$ of community $\mathcal{C}$, as a suitable DPDL formula Φ. Intuitively, for each service $A_i, i = 0 \ldots n$, involved in the composition, Φ encodes (i) its current state, and in particular whether A_i is in a final state, and (ii) the transitions that A_i can and cannot perform, and in particular which component service(s) performed a transition. Additionally, Φ captures the following constraints: (i) initially all services are in their initial state, (ii) at each step at least one of the component FSM has moved, (iii) when the desired service is in a final state also all component services must be in a final state.

The following results hold [7, 6]:

1. From the tree model property, the DPDL formula Φ is satisfiable if and only if there exists a composition of A_0 wrt $A_1, \ldots, A_n$.
2. From the small model property, if there exists a composition of A_0 wrt $A_1, \ldots, A_n$, then there exists one which is a MFSM of size which is at most exponential in the size of the schemas of $A_0, A_1, \ldots, A_n$.
3. From the EXPTIME-completeness of satisfiability in DPDL and from point 1 above, checking the existence of a service composition can be done in EXPTIME.

As an example, we can encode in a DPDL formula ϕ both the client specification shown in Figure 1(b) and the services in the community of Figure 3. Then we can use a DPDL tableaux algorithm to verify the satisfiability of ϕ. Such an algorithm returns a model that corresponds to the composition shown in Figure 4 (cf. [6]).

4   The Service Composition Tool $\mathcal{ESC}$

In this section we discuss the prototype tool $\mathcal{ESC}$ that we developed to compute automatic service composition in our framework.

Figure 5 shows the high level architecture for $\mathcal{ESC}$. We assume to have a repository of services, where each service is specified in terms of both its static

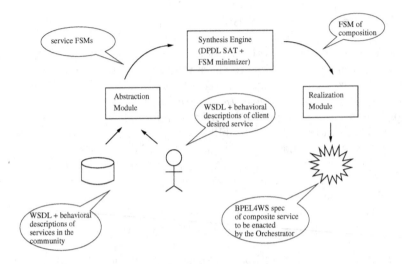

Fig. 5. The Service Composition Architecture

interface, through a WSDL document, and its behavioral description, which can be expressed in any language that allows to express a finite state machine (e.g., Web Service Conversation Language [13], Web Service Transition Language [9], BPEL4WS [2], etc.). The repository implements the community of services and can be seen, therefore, as an advanced version of UDDI. The client specifies his desired service in terms of a WSDL document and of its behavioral description, again expressed using one of the language mentioned before[10]. Both the services in the repository and the client desired service are then abstracted into the corresponding FMS (**Abstraction Module**). The **Synthesis Engine** is the core module of $\mathcal{ESC}$. It takes in input such FSMs, processes them according to our composition technique and produces in output the MFSM of the composite service, where each action is annotated with (the identifier of) the component service(s) that executes it. Finally, such abstract version of the composite service is realized into a BPEL4WS specification (**Realization Module**), that can be executed by an orchestration engine, i.e., a software module that suitably coordinates the execution of the component e-Services participating to the composition.

The implementation of the **Abstraction Module** depends on which language is used to represent the behavioral description of services.[11] In the current prototype we have considered Web Service Transition Language, which can be translated into FSMs [9]. Therefore, (for the moment) the abstraction module can deal only with it.

---

[10] We assume that the behavioral description of both the client specification and the services in the repository are expressed in the same language.

[11] In particular, the Abstraction Module is constituted by one submodule for each language used to specify a behavioral description as FSM. Such submodules can be easily plugged-in each time a new language is used.

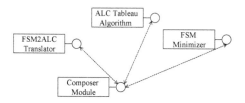

Fig. 6. Sub-modules of the `Synthesis Engine`

In the next subsections we will explain in detail the implementation of the `Synthesis Engine` and of the `Realization Module`.

4.1 Implementation of the Synthesis Engine Module

From a practical point of view, in order to actually build a finite state composition, we resort to Description Logics (DLs [3]), because of the well known correspondence between Propositional Dynamic Logic formulas (which DPDL belongs to) and DL knowledge bases. Tableaux algorithms for DLs have been widely studied in the literature, therefore, one can use current highly optimized DL-based systems [16, 14] to check the *existence* of service compositions. However, such the state-of-the-art DL reasoning systems cannot be used to *build* a finite state composition because they do not return a model. Therefore, we developed our $\mathcal{ESC}$ that, implementing a tableau algorithm for DL, builds a model (of the DL knowledge base that encodes the specific composition problem) which is a finite state composition. For our purpose the well-known $\mathcal{ALC}$ [3], equipped with the ability of expressing axioms, suffices.

The various functionalities of the `Synthesis Engine` are implemented into three Java sub-modules, as shown in Figure 6.

- The `FSM2ALC Translator` module takes in input the FSMs produced by the `Abstraction Module`, and translates them into an $\mathcal{ALC}$ knowledge base, following the encoding presented in [5].
- The $\mathcal{ALC}$ `Tableau Algorithm` module implements the standard tableau algorithm for $\mathcal{ALC}$ (cf., e.g., [10]): it verifies if the composition exists and if this is the case, it returns a model, which is a finite state machine.
- The `Minimizer` module minimizes the model, since it may contain states which are unreachable or unnecessary. Classical minimization techniques can be used, in particular, we implemented the *Implication Chart Method* [21]. The minimized FSM is then converted into a Mealy FSM, where each action is annotated with the service in the repository that executes it.

Since these three modules are in effect independent, they are wrapped into an additional module, the `Composer`, which also provides the user interface.

4.2   Implementation of the Realization Module

The technique for realizing an executable BPEL4WS file (i.e., an executable orchestration schema) starting from the automatically synthesized MFSM is as follows:

- Each transition in the MFSM corresponds to a BPEL4WS pattern consisting of *(i)* an `<onMessage>` operation (in order to wait for the input from the client of the composite service), *(ii)* followed by an invocation to the effective service (i.e., the deployed service that executes the operation), and then *(iii)* a final operation for returning the result to the client. Of course both before invoking the effective service and before returning the result, messages should be copied forth and back between the composite and the effective service. As an example, Figure 7 shows the BPEL4WS code corresponding to the MSFM transition for the `listen` operation relative to the MFSM of Figure 4.
- All the transitions originating from the same state are collected in a `<pick>` operation, having as many `<onMessage>` clauses as transitions originating from the state.
- The BPEL4WS file is built visiting the graph of the MFSM in depth, starting from the initial state and applying the previous rules, so that the nesting on `pick` and `sequence` operations reproduces the automata behavior. In Figure 8 it is shown the pseudo-code[12] of the whole BPEL4WS file obtained by the MFSM of Figure 4.

The BPEL4WS files thus produced can be deployed and executed onto standard BPEL4WS orchestration engines. In particular, we have tested such files using Collaxa BPEL Server 2.0[13].

5   Final Remarks and Future Work

In this paper we have presented $\mathcal{ESC}$, a prototype tool for automatic composition, which starting from a client specification and a set of available services, synthesizes a finite state composition.

We are currently extending our framework by allowing some advanced forms of non-determinism in the client specification and we are studying automatic composition techniques in this enhanced framework. In the future, we plan to produce a new version of our prototype tool that takes such extensions into account.

Finally, far-reaching future work may be identified along several directions. First of all, it could be interesting to study the situation when the available services export a partial description of their behavior, i.e., they are represented by non deterministic FSMs. This means that, a large (possibly infinite) number

---

[12] For sake of simplicity, we omit all BPEL4WS details and provide an intuitive, yet complete skeleton of the BPEL4WS file.

[13] cf. http://www.collaxa.com.

```xml
<?xml version="1.0" encoding="UTF-8"?>
<process ... >

    <partnerLinks>
    <!-- The 'client' role represents the requester of this service. It is used for callback.
         In our case it is the client of the composite service -->
            <partnerLink name="client"
                         partnerLinkType="tns:Transition"
                         myRole="MP3ServiceTypeProvider"
                         partnerRole="MP3ServiceTypeRequester"/>
            <partnerLink name="service"
                         partnerLinkType="nws:MP3CompositeService"
                         myRole="MP3ServiceTypeRequester"
                         partnerRole="MP3ServiceTypeProvider"/>
    </partnerLinks>

    <variables>
    <!-- Reference to the message passed as input during initiation -->
        <variable name="input" messageType="tns:listen_request"/>
        <!-- Reference to the message that will be sent back to the
             requestor during callback -->
            <variable name="output" messageType="tns:listen_response"/>
            <variable name="request" messageType="nws:listen_request"/>
            <variable name="response" messageType="nws:listen_response"/>
    </variables>

    <pick>
        <onMessage partnerLink="client"
                   portType="tns:MP3ServiceType"
                   operation="listen"
                   variable="input">
            <sequence>
                <assign>
                    <copy>
                        <from variable="input" part="selectedSong"/>
                        <to variable="request" part="selectedSong"/>
                    </copy>
                </assign>
                <invoke partnerLink="service"
                        portType="nws:MP3ServiceType"
                        operation="listen"
                        inputVariable="request"
                        outputVariable="response"/>
                <assign>
                    <copy>
                        <from variable="response" part="MP3FileURL"/>
                        <to variable="output" part="MP3FileURL"/>
                    </copy>
                </assign>
                <reply name="replyOutput"
                       partnerLink="client"
                       portType="tns:MP3ServiceType"
                       operation="listen"
                       variable="output"/>
                <!-- Other operations here for describing the next transitions -->
            </sequence>
        </onMessage>
        <onMessage>
        <!-- Other sequences here for describing the other possible transitions originating
             from the same state -->
        </onMessage>
    </pick>
</process>
```

Fig. 7. BPEL4WS code for the listen transition of the MFSM shown in Figure 4

```
<process>
    <pick>
        <onMessage="t">
            <sequence>
                <copy>...</copy>
                <invoke operation="t" on service A2 />
                <copy>...</copy>
                <reply ... />
                <pick>
                    <onMessage="l">
                        <sequence>
                            <copy>...</copy>
                            <invoke operation="l" on service A2 />
                            <copy>...</copy>
                            <reply ... />
                        </sequence>
                    </onMessage>
                </pick>
            </sequence>
        </onMessage>
        <onMessage="a">
            <sequence>
                <copy>...</copy>
                <invoke operation="a" on service A1 />
                <copy>...</copy>
                <reply ... />
                <pick>
                    <onMessage="l">
                        <sequence>
                            <copy>...</copy>
                            <invoke operation="l" on service A1 />
                            <copy>...</copy>
                            <reply ... />
                        </sequence>
                    </onMessage>
                </pick>
            </sequence>
        </onMessage>
    </pick>
<process>
```

Fig. 8. BPEL4WS pseudo-code for the MFSM shown in Figure 4

of complete description for services in the community exists that are coherent with each partial description. In such case, the orchestration schema that is to be synthesized should be coherent with all such possible complete descriptions. Therefore, computing composition in such a framework is intuitively much more difficult that in the framework presented here.

Also it is interesting to study how to add data in our framework and how this impacts the automatic service composition. In particular, it is worth studying how to introduce data in a way that the problem of automatic service composition, while exponential in the size of the service description, remains polynomial in the size of the data.

Finally, we foresee the validation of our approach and an engineered implementation of the tool in the context of the eG4M (e-Government for Mediterranean countries) project, in which the services offered by different Public Administrations spread all over some Mediterranean countries will be composed and orchestrated in order to offer value-added cooperative processes to citizens and enterprises.

Acknowledgements

This work has been supported by MIUR through the "FIRB 2001" project *MAIS* (http://www.mais-project.it, Workpackage 2), and "Societá dell'Informazione" subproject SP1 "Reti Internet: Efficienza, Integrazione e Sicurezza". It has been also supported by the European projects SEWASIE (IST-2001-34825), EU-PUBLI.com (IST-2001-35217) and INTEROP Network of Excellence (IST-508011).

The authors would like also to thank Alessandro Iuliani, for collaborating in the design and realization of the $\mathcal{ESC}$ tool, and Alessia Candido for her technical support with BPEL4WS.

References

1. G. Alonso, F. Casati, H. Kuno, and V. Machiraju. *Web Services. Concepts, Architectures and Applications*. Springer-Verlag, 2004.
2. T. Andrews, F. Curbera, H. Dholakia, Y. Goland, J. Klein, F. Leymann, K. Liu, D. Roller, D. Smith, S. Thatte, I. Trickovic, and S. Weerawarana. Business Process Execution Language for Web Services (Version 1.1). http://www-106.ibm.com/developerworks/library/ws-bpel/, May 2004.
3. F. Baader, D. Calvanese, D. McGuinness, D. Nardi, and P. F. Patel-Schneider, editors. *The Description Logic Handbook: Theory, Implementation and Applications*. Cambridge University Press, 2003.
4. C. Batini and M. Mecella. Enabling Italian *e*-Government Through a Cooperative Architecture. *IEEE Computer*, 34(2), 2001.
5. D. Berardi, D. Calvanese, G. De Giacomo, M. Lenzerini, and M. Mecella. service Composition by Description Logic Based Reasoning. In *Proceedings of the Int. Workshop on Description Logics (DL03)*, Rome, Italy 2003.
6. D. Berardi, D. Calvanese, G. D. Giacomo, M. Lenzerini, and M. Mecella. Automatic composition of *e*-services. Technical Report 22-03, Dipartimento di Informatica e Sistemistica, Università di Roma "La Sapienza", 2003.
7. D. Berardi, D. Calvanese, G. D. Giacomo, M. Lenzerini, and M. Mecella. Automatic composition of *e*-services that export their behavior. In *Proc. of the 1st Int. Conf. on Service Oriented Computing (ICSOC2003)*, 2003.
8. D. Berardi, D. Calvanese, G. D. Giacomo, M. Lenzerini, and M. Mecella. A foundational vision of *e*-services. In *Proc. of the CAiSE 2003 Workshop on Web Services, e-Business, and the Semantic Web (WES 2003)*, 2003.
9. D. Berardi, F. De Rosa, L. De Santis, and M. Mecella. Finite State Automata as Conceptual Model for e-Services. In *Journal of Integrated Design and Process Science*, 2004. To appear.
10. M. Buchheit, F. M. Donini, and A. Schaerf. Decidable reasoning in terminological knowledge representation systems. *J. of Artificial Intelligence Research*, 1:109–138, 1993.
11. T. Bultan, X. Fu, R. Hull, and J. Su. Conversation Specification: A New Approach to Design and Analysis of E-Service Composition. In *Proceedings of the WWW 2003 Conference*, Budapest, Hungary, 2003.
12. E. Colombo, C. Francalanci, B. Pernici, P. Plebani, M. Mecella, V. De Antonellis, and M. Melchiori. Cooperative Information Systems in Virtual Districts: the VISPO Approach. *IEEE Data Engineering Bulletin*, 25(4), 2002.

13. A. K. H. Kuno, M. Lemon and D. Beringer. Conversations + Interfaces = Business Logic. In *Proceedings of the 2nd VLDB International Workshop on Technologies for e-Services (VLDB-TES 2001)*, Rome, Italy, 2001.

14. V. Haarslev and R. Möller. RACER System Description. In *Proc. of IJCAR 2001*, volume 2083 of *LNAI*, pages 701–705. Springer-Verlag, 2001.

15. D. Harel, D. Kozen, and J. Tiuryn. *Dynamic Logic*. The MIT Press, 2000.

16. I. Horrocks. The FaCT System. In H. de Swart, editor, *Proc. of TABLEAUX'98*, volume 1397 of *LNAI*, pages 307–312. Springer-Verlag, 1998.

17. R. Hull, M. Benedikt, V. Christophides, and J. Su. E-Services: A Look Behind the Curtain. In *Proceedings of the PODS 2003 Conference*, San Diego, CA, USA, 2003.

18. S. McIlraith, T. Son, and H. Zeng. Semantic web services. *IEEE Intelligent Systems*, 16(2), 2001.

19. M. Papazoglou and D. Georgakopoulos. Service Oriented Computing (special issue). *Communications of the ACM*, 46(10), October 2003.

20. M. Pistore, F. Barbon, P. Bertoli, D. Shaparau, and P. Traverso. Planning and Monitoring Web Service Composition. In *Proc. of ICAPS Workshop on Planning for Web and Grid Service (P4WGS 2004)*, 2004.

21. R.H. Katz. *Contemporany Logic Design*. Benjamin Commings/Addison Wesley Publishing Company, 1993.

Dynamically Self-Organized Service Composition in Wireless Ad Hoc Networks

Qing Zhang[1], Huiqiong Chen[2], Yijun Yu[3], Zhipeng Xie[1], and Baile Shi[1]

[1] Department of Computing and Information Technology, Fudan University
qzhang79@yahoo.com, {xiezp, bshi}@fudan.edu.cn
[2] Faculty of Computer Science, Dalhousie University
hchen3@dal.ca
[3] Department of Computer Science, University of Toronto
yijun@cs.toronto.edu

Abstract. Service composition is a powerful tool to create new services rapidly by reusing existing ones. Previous research mainly focuses on the wired infrastructure-based environment. With the developments in mobile devices and wireless communication technology in recent years, mobile ad hoc network has received an increasing attention as a new communication paradigm. However, the existing service composition techniques do not work any longer in an ad hoc environment. In this paper, we present the service composition problem in wireless ad hoc network with full consideration of the characteristics of an ad hoc environment. To solve this problem, we develop two service composition routing algorithms, Simple Broadcasting Service Composition and Behavior Evolution Service Composition. The main contribution of our algorithms is that the whole process of service composition is done by the cooperation of nodes on-the-fly instead of a centralized broker to meet the peculiarity of ad hoc networks. Finally, we describe an initial implementation architecture for service composition in wireless ad hoc networks.

1 Introduction

Service composition refers to the technique of composing several existing services into a meaningful, richer service to meet the changing requirements of users. A service can be any functional program that produces corresponding output with appropriate input. We can view a complex and dynamic task as the composition of several basic sub-tasks, which can be completed by the cooperation of several simpler services.

The past research in service composition [1, 4, 8] mainly focuses on composing various services that are available over the fixed network infrastructure, where the physical location of a service does not need too much care. Existing service composition systems like eFlow [1], Ninja [4], and CMI [8] primarily rely on a centralized composition engine to carry out the discovery, integration and composition of web-based e-services [2].

M.-C. Shan et al. (Eds.): TES 2004, LNCS 3324, pp. 95–106, 2005.

Recent years have seen an increasing use of wireless mobile devices like mobile phones and PDAs. The advances in mobile devices and wireless communication technology have enabled a new communication paradigm: mobile ad hoc network. A mobile ad hoc network is a multi-hop, self-organized wireless network where the communication of mobile nodes can be done without the support of any fixed infrastructure. In this new networking environment, a mobile node acts as service provider and consumer at the same time. We can fulfill a complex task by the cooperation of services available in our vicinity.

Here, let's imagine a scenario in the future. Tom is chatting with his friend in an office when his Bluetooth-enabled cell phone beeps, indicating that he receives an email with attachment. The email says that the attachment contains some beautiful photos. Tom wants to browse the photos but unfortunately the attachment is a zipped file and his cell phone has no programs to unzip the attachment. Tom goes into a lab nearby with the cell phone. Suppose a Bluetooth-enabled laptop and a Bluetooth-enabled printer are provided in the lab. Tom can send the attachment to the laptop and unzip the attachment to browse the photos. Similarly, Tom can use the laptop to discover the printer and print the photos out.

From the example above, it can been seen that service composition in ad hoc network works in a highly dynamic manner. A mobile node tries to complete a task that it cannot accomplish by discovering and integrating the services in its vicinity. The centralized architecture of service composition in fixed network does not work any longer in ad hoc environment because we cannot find a stable centralized node which is omniscient to know everything, then carries out the composition of services in such a dynamic environment.

To the best of our knowledge, little work has addressed the problem associated with service composition in an ad hoc environment except [2]. In [2], the authors presented a distributed, de-centralized and fault-tolerant design architecture for reactive service composition in an ad hoc environment. They introduced two reactive techniques, "Dynamic Brokerage Selection" and "Distributed Brokerage technique", to accomplish service composition in dynamic environments. The center concept to their approach is that any node can act as the broker, which makes the design immune to single point of failure [2].

However, [2]'s approach relies too much on the node acting as broker. The selected broker is responsible for the whole composition process for a certain request. It first splits a task into sub-tasks to determine the necessary services, then discovers, integrates and executes these services to complete the task. In our opinion, the capability of a mobile node is limited while the requirement of user is changeable. How to execute a task may not be determined beforehand but depend on current situation, especially in a dynamic ad hoc environment.

In this paper, we deal with the problem of service composition in wireless ad hoc networks. We give a formal definition of this problem first, which fully considers the characteristics of an ad hoc environment (i.e. the network is changing dynamically and each mobile node only has knowledge of its current vicinity). Then we develop two routing algorithms called Simple Broadcasting Service Composition and Behavior Evolution Service Composition to solve this problem.

The key idea of our algorithms is that the whole process of service composition is dynamic and self-organized, which is done by the cooperation of nodes on-the-fly instead of a centralized broker.

The rest of the paper is organized as follows. Section 2 gives some basic definitions related to the problem. Section 3 presents the service composition problem in wireless ad hoc networks formally. Section 4 proposes two service composition routing algorithms to solve this problem. Section 5 presents an initial implementation architecture. We conclude our work in Sect. 6.

2   Definition

Before going into the service composition problem in ad hoc networks, we give some definitions.

Definition 1. *A service S can be specified as a tuple $\langle ID, I, O, F, C \rangle$, where:*
ID is the unique identification of service S;
I is the input pattern of service S, any input i that matches I can be an used as input of S;
O is the output pattern of service S, any output o that matches O can be used as an output of S;
F is the function that service S provides, an input i can be converted to o by F;
C is the cost for performing service S. It is related to multi-factors, such as the time spending in converting, the stability and battery power of current node where S locates in etc.

Every mobile device can provide one or more services $S_1, S_2, \cdots, S_n$, which can be defined as a service set $\{S_1, S_2, \cdots, S_n\}$.

Definition 2. *Given two services, $S_1 = \langle ID_1, I_1, O_1, F_1, C_1 \rangle$ and $S_2 = \langle ID_2, I_2, O_2, F_2, C_2 \rangle$, we say that S_1 and S_2 are k-hops range neighbors, if the mobile node N_1 where S_1 locates is no more than k-hops ($k \geq 1$) away from the mobile node N_2 where S_2 locate, or they are the same.*

Definition 3. *Assume that two services, $S_1 = \langle ID_1, I_1, O_1, F_1, C_1 \rangle$, $S_2 = \langle ID_2, I_2, O_2, F_2, C_2 \rangle$, are k-hops range neighbors, and o_1 is an output of S_1. If the whole or part of o_1 can be used as the input of S_2, we say that S_1 and S_2 can be completely composed in k-hops range, and the direction of the composition is from S_1 to S_2. It can be denoted as $S_1 \xrightarrow{k} S_2$, or $i_1 S_1 o_1 S_2 o_2$ in detail, where o_2 is the output of S_2 that takes o_1 as input.*

Definition 4. *Assume that two services, $S_1 = \langle ID_1, I_1, O_1, F_1, C_1 \rangle$, $S_2 = \langle ID_2, I_2, O_2, F_2, C_2 \rangle$, are k-hops range neighbors, and o_1 is an output of S_1. If the whole or part of o_1 can be used as part of the input of S_2, we say that S_1 and S_2 can be partially composed in k-hops range, and the direction of the composition is from S_1 to S_2. It can be denoted as $S_1 \xrightarrow{k} S_2$.*

We say S_1 and S_2 can be *composed in k-hops range* if they are completely composed or partially composed in k-hops range.

Definition 5. *Given two services,* $S_1 = \langle ID_1, I_1, O_1, F_1, C_1 \rangle$ *and* $S_2 = \langle ID_2, I_2, O_2, F_2, C_2 \rangle$. *If* S_1 *and* S_2 *can be performed independently, we say that* S_1 *and* S_2 *are independent, and denote it as* $S_1 \parallel S_2$.

From Definition 4 and Definition 5, we can get that:

Given services $S_1 = \langle ID_1, I_1, O_1, F_1, C_1 \rangle$, $S_2 = \langle ID_2, I_2, O_2, F_2, C_2 \rangle$ and $S_3 = \langle ID_3, I_3, O_3, F_3, C_3 \rangle$, $S_1 \parallel S_2$, $S_1 \xrightarrow{k} S_3$, $S_2 \xrightarrow{k} S_3$, o_1 is an output of S_1, and o_2 is an output of S_2. The union of o_1 and o_2 is denoted as $o_1 \cup o_2$. If the whole or part of $o_1 \cup o_2$ can be used as part of the input of S_3, we say that the *combination* of S_1 and S_2 can be completely composed with S_3 in k-hops range, and denote it as $(S_1 \cup S_2) \xrightarrow{k} S_3$.

3   Problem Statement

In this section, we present the service composition problem in wireless ad hoc networks.

In a dynamic ad hoc environment, there are n mobile nodes, each of which has a unique identification from id_1 to id_n. Every node id_m ($1 \le m \le n$) has p_m services from $S_{m,1}$ to S_{m,p_m}, and id_m only has knowledge (the services provided by node) of itself and nodes within k-hops range. Now some node id_s, which we call the *task initiator*, starts a task $t\langle tid, I(i), O \rangle$, where tid is the unique identification of t, i is the input of t that matches pattern I, and O is the output pattern which can be acquired when t is completed. Determine the *flow* to complete task t under the following restriction and then finish the task.

- The maximal number of hops should be less than a constant H;
- The time taken by the search process should be less than a constant T.

Our definition of problem has following features:

1. It is a decentralized one. There does not exist any centralized broker, which carries out the service composition process. Every node only has knowledge of itself and its current vicinity.
2. For every task t, only an input i and the output pattern O need to be provided. The services needed to complete t and the order of services are determined on-the-fly by the cooperation of nodes, which can satisfy the changeable requirements of users.
3. There may exist several flows to complete task t because only i and O are provided. This gives us the chance to select a best one as the solution of the task.

4   Service Composition Routing Algorithms

In this section, we introduce two routing algorithms for service composition in ad hoc environment: *Simple Broadcasting Service Composition* and *Behavior Evolution Service Composition*. As we have pointed out, in [2]'s approach, the node acting as broker is responsible for the whole composition process for a certain request, it relies too much on the selected broker even if any node can act as this broker. In our algorithms, the whole process of service composition is done on-the-fly by the cooperation of nodes instead of a centralized broker, which makes the key of our two algorithms. For the simplicity of description, we confine our discussion in 1-hop range and omit the declaration of "in 1-hop range" in the rest of our paper. The solution to k-hop range service composition problem can be inferred from the solution to 1-hop range service composition problem. The method is that each node finds its k-hop range neighbors and treats them the same way as it does to the 1-hop range neighbors.

4.1   Simple Broadcasting Service Composition

Overview. The Simple Broadcasting Service Composition method consists of two mechanisms: *Service Composition Flow Discovery* and *Service Composition Fault Recovery.*

To complete a task, the task initiator should discover the services needed to complete a task and decide the order of services, which are done by the Service Composition Flow Discovery procedure. In this procedure, beginning from the task initiator, a node that has service to continue task but cannot get the desired output pattern simply broadcasts its output pattern to its neighbors as input pattern, seeking the next node that has service to continue task. This process repeats till the desired output pattern is acquired, then the flow to carry out the task can be constructed and the task is executed according to the sequence of nodes in the flow.

When a task is being executed, the task initiator should monitor the progress of execution, which is done by the Service Composition Fault Recovery procedure. Wireless ad hoc networks are less stable environment than wired networks, where the nodes move frequently, and the power in nodes may sometimes become insufficient. The established flow is prone to be broken. To recover from faults occurred in the process of task execution, We adopt a fault recovery mechanism.

Service Composition Flow Discovery. We view the service composition flow as a DAG (directed acyclic graph) made up of nodes and edges. Figure 1 illustrates an example of the flow. A *flow node* is a service in a mobile node that can continue a task, which is denoted as the pair ⟨node, service⟩. A *flow edge* represents the two connected services can be composed. For example, in Fig.1, flow node 0 is a service 0 in mobile node 0; the edge between flow node 1 and 4 represents service 1 in mobile node 1 and service 4 in mobile node 3 can be completely composed, and the edge between flow node 3 and 6 represents service 3 in mobile node 2 and service 6 in mobile node 5 can be partially composed.

The graph starts from the flow node that can accept task input (*flow start node*), which is flow node 0 in Fig.1, and ends at the flow node that can generate output matching the desired task output pattern (*flow end node*), which is flow node 7 in Fig.1. In the flow, there may exist flow nodes which accept more than one input, such as flow node 6, which means the combination of some services can be completely composed with a service. For example, in Fig.1, the combination of service 3 in mobile node 2 and service 5 in mobile node 4 can be completely composed with service 6 in mobile node 5. We call this kind of node *flow key node*. A route from one flow key node (or flow start node) to next flow key node (or flow end node) is called *path*. In Fig.1, there are three paths, they are $0 \rightarrow 1 \rightarrow 2 \rightarrow 3 \rightarrow 6$, $0 \rightarrow 1 \rightarrow 4 \rightarrow 5 \rightarrow 6$, and $6 \rightarrow 7$. A flow can be viewed as the composition of several paths.

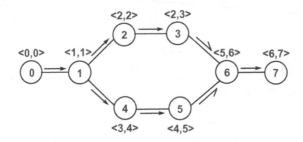

Fig. 1. A Flow Example

It is the duty of Service Composition Flow Discovery procedure to construct the flow. Now, a mobile node launch a task $t\langle tid, I(i), O\rangle$ at the time T_1. It broadcasts a *flow request* packet to itself and its neighbors. The packet contains the task related information $\langle tid, I, O, T_1 \rangle$ as well as a *path record*, which accumulates a sequence of $\langle id_k, S_t\langle ID_t, I_t, O_t, C_t\rangle\rangle$ in a path when the packet is propagated through the ad hoc network during the flow discovery. To prevent cycle in a flow, each mobile node maintains a list of $\langle task\ id,\ service\ id\rangle$ pairs when the node can continue a task through a service.

When a mobile node id_{k+1} receives a flow request packet, it processes the request as follow:

1. Check the launch time T_1 of task t. If the task is out of date, discard the flow request packet and do not process it further.
2. Otherwise, extract the last element in the path record, say $\langle id_k, S_t\langle ID_t, I_t, O_t, C_t\rangle\rangle$, then check the services id_{k+1} possesses to find whether there exists a service S_{t+1}, which can accept O_t as input, i.e. id_{k+1} tries to find a service S_{t+1}, which can be composed with the last service S_t in the path record. If there does not exist one, discard the flow request packet.
3. Otherwise, if S_t and S_{t+1} can be composed, but the pair $\langle tid, S_{t+1}\langle ID_{t+1}\rangle\rangle$ already exists in id_{k+1}'s list of $\langle task\ id,\ service\ id\rangle$ pairs, discard the flow request packet.

4. If S_t and S_{t+1} can be completely composed, append $\langle id_{k+1}, S_{t+1}\langle ID_{t+1}, I_{t+1}, O_{t+1}, C_{t+1}\rangle\rangle$ to the path record in the flow request packet.
5. If S_t and S_{t+1} can only be partially composed, wait for a given period T_w until the time is out, or id_{k+1} receives several other flow request packets, the combination of the last services in each path record can be completely composed with S_{t+1}. If the time is out, discard the flow request packet. Otherwise, $\langle id_{k+1}, S_{t+1}\langle ID_{t+1}\rangle\rangle$ can be a flow key node, append $\langle id_{k+1}, S_{t+1}\langle ID_{t+1}, I_{t+1}, O_{t+1}, C_{t+1}\rangle\rangle$ to the path record in each flow request packet, and cache the list of path records. Then rebuild a new path record in the flow request packet, which has one element $\langle id_{k+1}, S_{t+1}\langle ID_{t+1}, I_{t+1}, O_{t+1}, C_{t+1}\rangle\rangle$.
6. At this point, S_{t+1} can be completely composed with a service (or services). If the output pattern O_{t+1} of S_{t+1} is a superset of the desired output pattern O of task, it means that $\langle id_{k+1}, S_{t+1}\langle ID_{t+1}, I_{t+1}, O_{t+1}, C_{t+1}\rangle\rangle$ is the flow end node, and the paths to carry out task is successfully found, return a *flow reply* packet to the task initiator. Otherwise, continue the flow discovery process, and id_{k+1} re-broadcast the request to itself and its neighbors.

Now, the flow end node prepares to return a flow reply packet to the task initiator. The paths to construct the service composition flow are collected at the same time. The flow reply packet contains six fields: a unique *flow id* (*fid*), *task id* (*tid*), *current path record*, *current position* in path record, a list of *collected paths*, and the *maximal number of possible paths*. The list of collected paths is initialized with current path record, and the maximal number of possible paths is initialized to 1. To obtain all the paths for a flow *fid*, the task initiator maintains a list of *received paths* for each possible flow.

When a mobile node id_{k+1} receives a flow reply packet (or the packet is built by itself in case that id_{k+1} is the flow end node or a flow key node), it processes the request as following:

1. Provided current position in packet is $\langle id_{k+1}, S_{t+1}\langle ID_{t+1}\rangle\rangle$, track the elements of current path record from current position in a regressive direction.
2. If id_{k+1} can find an element $\langle id_k, S_t\langle ID_t, I_t, O_t, C_t\rangle\rangle$, and id_k and id_{k+1} are different mobile nodes, set current position to $\langle id_k, S_t\langle ID_t, I_t, O_t, C_t\rangle\rangle$, and send the flow reply packet to id_k.
3. Otherwise, it means that $\langle id_{k+1}, S_{t+1}\langle ID_{t+1}\rangle\rangle$ is the flow start node or a flow key node. If $\langle id_{k+1}, S_{t+1}\langle ID_{t+1}\rangle\rangle$ is a flow key node, id_{k+1} has the information of a list of path records for $\langle tid, S_{t+1}\langle ID_{t+1}\rangle\rangle$ which has been cached in the process of sending the flow request packet, add the number of path records to the maximal number of possible paths. For each path, rebuild a new flow reply packet, the new current path record is the corresponding path record, and the new list of collected paths is the original list of collected paths appended with the new current path record. Re-process the flow reply packet respectively.

4. Otherwise, $\langle id_{k+1}, S_{t+1} \langle ID_{t+1} \rangle \rangle$ is the flow start node, append the list of collected paths in the packet to the list of received paths for fid that id_{k+1} maintains, and remove the duplicated paths. If the number of received paths is equal to the maximal number of possible paths, it means that all paths to carry out task tid are reachable during the process of sending the flow reply packet, and we call this list of received paths is *valid*.

In a period of time T, the task initiator may obtain several valid lists of received paths, from which we can construct the flow to carry out the task. Compute the total cost for each flow, select the one which has the minimal cost as the flow to carry out the task, and then start executing task.

Service Composition Fault Recovery. During the execution of a task, the established flow has a high probability of being broken due to the nature of wireless networks. We need adopt a fault recovery mechanism to deal with it.

The simplest solution for this problem is to select an alternative flow, or restart the task. Since the task initiator perhaps has received several useful flow reply packets, and constructed more than one flow in the phase of Service Composition Flow Discovery, we can select another inferior flow to carry out the execution of task, or simply initiate a new Service Composition Flow Discovery phase if the selected flow through which the task is being executed is broken.

It is clear that the method above is inefficient because failures may occur with high probability, so we adopt a similar fault recovery mechanism as the one described in [2], which employs a checkpoint technique to guard against failures. After a flow node completes its subtask, it sends back its partial completion state and the checkpoint to the task initiator. The task initiator caches this partial result obtained so far. If a flow node fails, the task initiator can receive no more checkpoints. It then reconstructs the task that is still left to solve, which is treated as a new task, and restarts the new task.

4.2   Behavior Evolution Service Composition

The broadcasting method we describe above has an obvious fault: the heavy packet overhead. When there are rich resources in the vicinity for every mobile node, the packet overhead will be extremely heavy because every flow node tries its best to compose with its neighbor services with no selections.

To optimize the broadcasting method, we purpose a new *Behavior Evolution Service Composition* method based on it. In this method, taking full use of the experiences (expressed by rules) acquired from previous service composition process, every flow node tries to select a certain number of services from its neighbor services which it can be composed with in a probabilistic approach. Here *behavior* means the selection of services which a flow node can be composed with; *evolution* means that the selection process can be increasingly efficient with the accumulation of experiences.

This method consists of two mechanisms: *Rule Acquirement* and *Rule Utilization*.

Rule Acquirement. Every mobile node can accumulate the experiences acquired from the cooperation of mobile nodes to accomplish a task. When a task comes into the phase of being executed, let the flow for this task transferred with the execution of task. The flow has the information of the next services to be composed with for a certain output pattern. We treat this information as a sign of experiences.

We can express experiences by the form of rules. For a flow node, we generate rules according to the flow like this: the precondition of rule is the output pattern of a flow node which is reachable from current flow node, the postcondition of rule is the service(s) to be composed with:

$$S_t: \text{IF } O \text{ THEN } \{S_{t+1}\langle ID_{t+1}\rangle\}$$

It means that service S_t is composed with $\{S_{t+1}\langle ID_{t+1}\rangle\}$ to obtain the output pattern O according to current flow.

We measure a rule by two metrics: *confidence* and *age*. When a rule r is generated by a node id_k, we change the confidence and age of rules in id_k as following:

– If r does not exist in id_k, we set its confidence to 1. Otherwise, we increase its confidence by one. The confidence of other rules do not change.
– No matter r exists in id_k or not, we set its age to 1, and increase the age of other rules by one.

Each node id_k is given a limited space to store rules. If id_k has not enough space to store a new generated rule, we employ following rule replacement strategy: To all rules in id_k and the new generated rule, compute the ratio of confidence to age for each rule, and discard the rule which has the minimal ratio.

Rule Utilization. To utilize the rules we have got, we make some modifications to the phase of Service Composition Flow Discovery in broadcasting method. A mobile node id_k no longer broadcasts a flow request packet to its neighbors but sends the packet to a certain number of selected neighbors instead. To determine which neighbors to be selected, the mobile node broadcasts a *service query* packet first. The packet contains $\langle tid, id_k, S_t\langle ID_t, O_t\rangle\rangle$, which indicates the service in id_k to continue task.

When a mobile node id_{k+1} receives a service query packet, it processes the request as following:

1. id_{k+1} checks the services it possesses to find a service S_{t+1} which can be composed with S_t. If there does not exist one, discard the service query packet.
2. Otherwise, if S_t and S_{t+1} can be composed, but the pair $\langle tid, S_{t+1}\langle ID_{t+1}\rangle\rangle$ already exists in id_{k+1}'s list of $\langle taskid, serviceid\rangle$ pairs, discard the service query packet.
3. Otherwise, return a *service acknowledge* packet to id_k. The packet contains $\langle id_{k+1}, S_{t+1}\langle ID_{t+1}, O_{t+1}, C_{t+1}\rangle, M\rangle$, where M indicated whether id_k and id_{k+1} can be completely composed or not.

In a period of time T_q, id_k may has received several return service acknowledge packets. It is clear that a service can be selected if the output pattern of the service is a superset of the desired output pattern O of task. For the remaining services, we should make a selection from these services by utilizing the rules it accumulates. We make the following assumptions:

1. The number of services is m.
2. There are k special rules accumulated in id_k, each of these k rules has a precondition O (i.e. the output pattern of task), and a postcondition belonging to these m services. We denote these k rules from r_1 to r_k, and their corresponding confidence from b_1 to b_k.
3. There are m_1 services in all postcondition of these k rules, and m_2 ($m_2 = m - m_1$) services out of these k rules.

Try to select u_1 rules and corresponding services from these k rules, and select u_2 services from the m_2 services with consideration of the confidence of a rule and the cost for performing a service.

If $k \leq u_1$, use all k rules, $u_2 = u_2 + u_1 - k$.

If $m_2 \leq u_2$, use all m_2 services.

If $k > u_1$ or $m_2 > u_2$, we do the selection by using the probabilistic Monte Carlo method [7] as following:

- For each rule r_t: IF O THEN $\{S_{t_1}, S_{t_2}, ..., S_{t_l}\}$, we define the weight w_t of r_t as $b_t / \sum_{j=1}^{l} S_{t_j} \langle C_{t_j} \rangle$. Then the probability of selecting services $\{S_{t_1}, S_{t_2}, ..., S_{t_l}\}$ is $w_t / \sum_{i=1}^{k} w_i$.
- For each service S_t in m_2 services, the probability of S_t being selected is $(1/S_t \langle C_t \rangle) / \sum_{j=1}^{m_2} (1/S_j \langle C_j \rangle)$.

Now, id_k has got a certain number of selected services from the services gathered, it will send a new flow request packet to every corresponding mobile node. The new packet contains $\langle S_{t+1} \langle ID_{t+1} \rangle, M \rangle$ in addition. When a mobile node id_{k+1} receives the new flow request packet, it treat the request in a similar way as the process when id_{k+1} receives a flow request packet, except that it is unnecessary to do step 2 and 3 in the broadcasting method.

5   Implementation

To check the validity of our algorithms, we have implemented the basic mechanism of our two service composition routing algorithms in NS2 [3] with mobility and wireless extensions. NS2 is a discrete event network simulator developed by the University of California, Berkeley and the VINT project. Rice Monarch Project [9] has made substantial extensions to NS2 to allow ad hoc simulations.

For each mobile node, we use the architecture shown in Fig.2. It mainly includes following three modules:

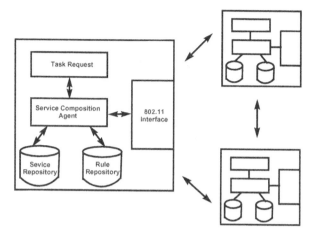

Fig. 2. Service Composition Architecture

Service Composition Agent: The service composition agent is the core of our architecture. It accepts the task requested by user and tries to find the service composition flow to execute task. We have completed an initial implementation of our two service composition methods under the framework of NS2. We also carried out various experiments to compare these two methods by measuring the packet overhead and task completion ratio. As we have expected, the Behavior Evolution Service Composition method has higher performance than the Simple Broadcasting Service Composition method, which can reduce the packet overhead greatly with little loss of the task completion ratio.

Service Repository: The service repository module describes the basic information (input pattern, output pattern and function) of services a node has and the relationship between services. Currently we simplified the representation of service information by text string. It will be our future work to use a common language like DAML-S (the DARPA Agent Markup Language for Service) [10] to describe the information of a service.

Rule Repository: The rule repository module stores the rules accumulated in the process of service composition. These rules can be used to direct the future service composition process. Taking advantage of this feature, if we already have the knowledge of an existing service composition flow, we can express the flow by the form of rules with high confidence, and deposit these rules in the nodes which have corresponding services. It will also be useful to collect and analysis the rules accumulated in several nodes. A rule has high confidence means the services in the rule are composed frequently. We can select the rules with high confidence, combine these services in rules that can be composed into a new service composition flow, and treat this flow as a new service because the services in the flow are used together frequently. It will bring the emergence of a new service [5].

6   Conclusions

In this paper, we cope with the service composition problem for ad hoc networks. By considering the features of ad hoc environment, we present the problem and develop two service composition routing algorithms: Simple Broadcasting Service Composition and Behavior Evolution Service Composition. In the broadcasting method, a node which can continue a service composition task simply broadcasts its intermediate output to its neighbors till the desired output is acquired. The packet overhead of broadcasting method is extremely heavy. To optimize it, we propose the behavior evolution method, which takes use of experiences accumulated in previous service composition process. Finally, we present an initial implementation architecture in NS2. It will be our future work to perfect our system and make it work in a real world.

References

1. F. Casati, S. Ilnicki, L. Jin, V. Krishnamoorthy, and M. C. Shan. Adaptive and dynamic service composition in eflow. *Proceedings of the International Conference on Advanced Information Systems Engineering*, June 2000.
2. D. Chakraborty, F. Perich, A. Joshi, T. Finin, and Y. Yesha. A reactive service composition architecture for pervasive computing environment. *7th Personal Wireless Communications Conference*, 2002.
3. K. Fall and K. Varadhan. The ns Manual. *The VINT Project*, September 2003.
4. S. D. Gribble, M. Welsh, R. V. Behren, E. A. Brewer, etc. The Ninja Architecture for Robust Internet-Scale Systems and Services. *IEEE Computer Networks Special Issue on Pervasive Computing*, March 2001, Vol 35, No. 4.
5. T. Itao, S. Tanaka, T. Suda, and T. Aoyama. Adaptive Creation of Mobile Network Applications in the Jack-in-the-Net Architecture. *Wireless Networks, the Journal of Mobile Communication, Computation and Information*, vol. 10, issue 3, pp.287-299, May 2004.
6. D. Johnson and D. Maltz. Dynamic source routing in ad hoc wireless networks. *Mobile Computing*, Kluwer Academic Publishers, 1996.
7. J. M. Pollard. Monte Carlo Methods for Index Computation (mod p). *Mathematics of Computation*, July 1978.
8. H. Schuster, D. Georgakopoulos, A. Cichocki, and D. Baker. Modeling and composing service-based and reference process-based multi-enterprise processes. *Proceedings of the International Conference on Advanced Information Systems Engineering*, June 2000.
9. The Rice University Monarch Project. World Wide Web, http://www.monarch.cs.rice.edu/.
10. DARPA Agent Markup Language for Services. World Wide Web, http://www.daml.org/services/.

Designing Workflow Views with Flows for Large-Scale Business-to-Business Information Systems

Dickson K.W. Chiu[1], Zhe Shan[2], Patrick C. K. Hung[3], and Qing Li[2]

[1] Dickson Computer Systems, 7A Victory Avenue 4th floor,
Homantin, Kln, Hong Kong
dicksonchiu@ieee.orga
[2] Department of Computer Engineering and Information Technology,
City University of Hong Kong
{zshan0, itqli}@cityu.edu.hk
[3] Faculty of Business and Information Technology,
University of Ontario Institute of Technology, Canada
patrick.hung@uoit.ca

Abstract. Workflow technology has recently been employed as a framework for implementing large-scale business-to-business (B2B) information systems over the Internet. This typically requires collaborative enactment of complex workflows across multiple organizations. To tackle the complex of these cross-organizational interactions, we propose a methodology to break down workflow requirements into five types of elementary flows: control, data, semantics, exception, and security flows. Then, we can determine the subset of each of five types of flows necessary for the interactions with each type of business partners. These five subsets, namely, flow views, constitute a workflow view, based on which interactions can be systematically designed and managed. We further illustrate how these flows can be implemented with various contemporary Web services standard technologies.

1 Introduction

An important challenge in B2B information system is interaction. Interaction is defined as consisting of interoperation and integration with both internal and external enterprise applications. This has been a central concern because these applications are composed of autonomous, heterogeneous, and distributed business processes.

The problem is prominent in large-scale B2B information systems, where interactions are complex and varies across different types of business partners. We have proposed a workflow view based approach to address this problem in our previous work [1-3]. A workflow view is a structurally correct subset of a workflow [1]. The use of workflow views facilitates sophisticated interactions among workflow management systems (WFMSs) and allows these interactions to inter-operate in a gray box mode (that is, they can access each other's internal information to some extent). In addition, workflow views are useful in providing access to business processes for external customers or users, including B2C e-commerce. The artifact of workflow views is therefore a handy mechanism to enact and enforce cross-organizational inter-operability in e-services.

M.-C. Shan et al. (Eds.): TES 2004, LNCS 3324, pp. 107–121, 2005.

However, the design of workflow views is still not obviously in large-scale information systems. To tackle this problem, we now propose using the concept of flows
as shown in Fig. 1, a conceptual model in Unified Model Language (UML) class
diagram. We partition the interactions into five types of requirements: control, data,
semantics, exception, and security. Such interactions involve the communication of
events, which is an atomic occurrence of something interesting to the system itself or
user applications. A flow is a directed relationship that transmits an event from a
source activity to a sink activity between the business partners. The corresponding
flows to the requirements are therefore control flow, data flow, semantics flow, exception flow, and security flow, respectively.

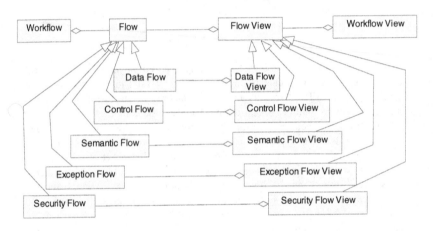

Fig. 1. Conceptual Model of Workflow Views and Flows

After the partitioning, we can determine the subset of each of five types of flows
necessary for the interactions with each type of business partners. These five subsets
are therefore, control flow view, data flow view, semantics flow view, exception flow
view, and security flow view respectively. These five flow views collectively constitute a workflow view based on which interactions can be systematically designed and
managed.

We propose the formulation of flow in a loosely coupled Web services environment because current trends in information and communication technology accelerate
the widespread use of Web services in information systems. In this paper, a Web
service refers to an autonomous unit of application logic that provides some information processing resources to other applications through the Internet from a service
provider. We employ various contemporary Web services technologies for adding
advanced control to simple procedure invocations. We define semantics references in
Business Process Execution Language for Web Services (BPEL4WS) [4] by using
the Web Ontology Language (OWL) [5] to provide explicit meaning to information
available on the Web for automatic process and information integration. In the aspect
of exception handling, we link the proposed exception-handling assertions in

BPEL4WS to SOAP-fault implementations and examine some typical use cases of exception. Further to increase the flexibility and alternatives in handling exceptions, we also discuss how to employ Semantic Web technologies in handling exceptions in addition to human intervention support.

The rest of the paper is organized as follows. Section 2 introduces semantics, control, and data flows while section 3 describes security and exception flows. Section 4 gives an example how semantics can help exception handling. Section 5 summarizes the formulation of workflow views based on various flows with an example. Section 6 reviews background and related work. We conclude the paper with our plans for further research.

2 Semantics, Control, and Data Flows

To illustrate our framework, we present a motivating example of intelligence information integration in an investigation of a suspect. We use this example instead of a medical one as there are more semantic issues and there are more alternatives ways for investigating people's data, which requires often approval. For example, a detective investigates a suspect by inspecting an integrated view of records (e.g., criminal records, border control, and bank transactions) sourced from different government

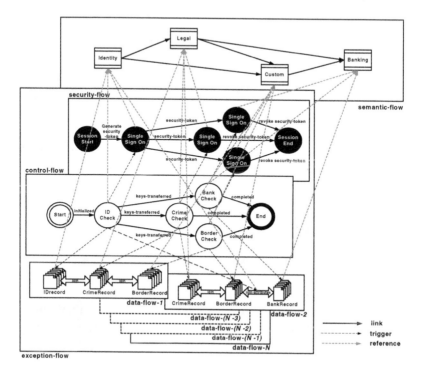

Fig. 2. An Example of Control-flow, Data-flows, Semantics-flow, and Security-flow

and commercial organizations. In particular, bank transactions within one month before and after a trip above a certain threshold amount are retrieved. To illustrate, a sample WII schema with the following relations and attributes (in parentheses) is shown as follows:

- IDrecord (id-no, tax-file-no, name, sex, date-of-birth, area-code, phone-no, address, postal-code)
- CrimeRecord (id-no, crime-description, sentence, day-of-event)
- BorderRecord (id-no, entry-or-exit, place, vehicle, day-of-event)
- BankRecord (tax-file-no, bank-no, account-no, transaction, amount, balance, day-of-event)

Fig. 2 illustrates the flow technologies to manage and monitor the control-flows, data-flows, and semantics-flows in a large-scale information system. Flow technology is becoming an integral part of modern programming models [6]. Each flow is separated and depicted in the context of a multi-layer framework. This is also called flow independency. The separation of flows results in increased flexibility of information Web services in executing workflows. Thus the workflow modelers can easily change or update the information integration plans for different situations.

Each service provider provides Web services at the service layer and BPEL4WS orchestrates them together in order to achieve integrated plans. Fig. 2 describes workflow that all the activities (in circle) are performed for retrieving the datasets from various databases (labeled by the activity's name), and are coordinated by a set of events (in single arrow lines). Each activity is assigned to a Web service for execution by a matchmaking process. In particular, each activity has to obtain a read-access approval from each data custodian and data service provider.

Semantics flow defines the semantic relationship among the information which will be used in the execution of the workflows. Although via applying the Semantic Web technologies the defined ontology has internally define the relationship of information semantics, the semantics flow abstract the main concepts and describe their dependence in a more clear way.   The data schema can be represented in OWL as ontology. Based on it, we propose the semantic-referencing assertions in BPEL4WS for supporting semantics flows in large scale information systems.

Control flow specifies the order of activities which will be conducted in the workflow. A control flow shows the picture of business logic in a workflow process.

Data flow defines the flow of specific data or dataset through a workflow. In simple workflows which only involve few data, its data flows are almost same as the control flow. But, large-scale workflow system deals with many data in parallel. Its control flow cannot show the action sequence of data in a clear way. Hence, we use data flow to clarify the vision.

The interactions among control flow, data flow, and semantics flow are triggered by external events (in dashed arrow lines). These external events contain the datasets generated from the activities. Referring to Fig. 2, there is a set of N data-views (N is a cardinal number) that are performed with the control-view. Each data-flow is also assigned to an information Web service for execution by a matchmaking process if necessary. In the context of BPEL4WS, we propose new data-integration assertions named <integrate>, <dataset> and <dataLinkage> for generating the data-flows. Referring to Fig. 2, both the control flow and data flow(s) reference to the relevant ontology described in the semantics flow. In a general case, the workflow is ended once all the control flow and data flow(s) are completed successfully.

3 Security and Exception Flows

We propose to manage, store, and represent an user's access control information as a security token in the context of WS-Security [7], which describes and provides protection enhancements to SOAP messaging to provide quality of protection through message integrity, message confidentiality, and single message authentication.

Based on the security token define in the SOAP header, we propose security view assertions in BPEL4WS as <sessionStart/>, <clearance/>, <securityToken/>, <token-Type/> and <sessionEnd/>. The <sessionStart/> assertion is used to identify the time when the user's security token is generated by the information system, and the <sessionEnd/> assertion is used to identify the time when the user's security token should be revoked. The security flow is orchestrated with the control flow. As such, authentication is not only based on the "Username" and "Password" but also other information such as the "SubjectName" and "SubjectLocation". Each of the activity can define whether the security clearance assertion <clearance/> is required and the details such as the type of security token <securityToken/> and <tokenType/>. For the tokens, Security Assertions Markup Language (SAML) [8] is used to define such authentication and authorization decisions. SAML is an XML-based framework for exchanging security credentials in the form of assertions about subjects.

In general, there are two types of exceptions in the proposed conceptual workflow model: expected and unexpected exceptions. Excepted exceptions are predicable deviations from the normal behavior of the workflow. In our proposed workflow model, there are five categories of expected exceptions. Control exceptions are raised in correspondence to control-flows such as start or completion of activities. Data exceptions are raised in correspondence to data-flows such as data integration processes. Temporal exceptions are raised in correspondence to both control-flows and data-flows such as the occurrence of a given timestamp or a pre-defined interval elapsed. External exceptions are raised in correspondence to control-flows and data-flows explicitly notified by external services such as system failures. Security exceptions are raised in correspondence to access control or security violations.

External and temporal exceptions are in general asynchronous, but control, data, and security exceptions occur synchronously with activity executions. In our proposed workflow model, unexpected exceptions mainly correspond to mismatches between an activity specification and its execution. In many cases, human intervention is a mechanism for handling unexpected exceptions.   Activities failure is defined as one or more activities have failed or are unavailable due to the context of activities execution. In generation, there are three common exception-handling procedures [9], namely, Remedy, Forward Recovery, Backward Recovery.

Anything that has an algorithmic flow also has a pervasive exception-handling need [10]. Exception flows are often asynchronous with respect to the control flows, data flows, and security flows, both in their raising and in their handling. Fig. 3 describes our proposed exception-handling approach in the different levels protocols (i.e., BPEL4WS and SOAP). Once a control, data, or security exception is raised, the corresponding Web service will generate a SOAP fault message as an exception event to trigger the exception-flows.

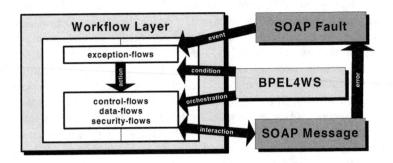

Fig. 3. Proposed Exception-handling Approach

We propose some new assertions to describe the exception-handling procedures in BPEL4WS. These assertions integrate with the exception-handling procedure specified by the conditions in BPEL4WS, so that appropriate actions can be taken in the context of control flows, data flows, and security flows.

Referring to the scenario of security-flows, there are two circumstances in which a security exception can occur: activity-specific or cross-activity. An activity-specific exception only affects exactly one activity, but a cross-activity exception may affect more than one activity.

In the context of BPEL4WS, we propose new exception-handling assertion named <exceptionHandling>, <event>, <condition>, and <action> for generating the data flows. Moreover, the proposed conceptual workflow model requires a termination mechanism to prevent exceptions trigger each other indefinitely. In the worse case, if the workflow designer cannot find any feasible exception-handling procedure, the problematic activity has to abort and so the user request has to abort as well. This is known as failure determination and is an undesirable situation in workflow execution. In this case, we propose new exception-handling assertion named <exceptionHandlingDefault> for specifying the abort action if none of the rules can handle the exception.

4   Exception Handling with Semantic Assistance

We further demonstrate the feasibility and advantages of employing Semantic Web technologies to assist in exception handling with two cases in security flow for remedy approach and forward recovery, respectively.

Imagine that there is a security exception occurred at the "ID Check" activity in the security-flow. The Web service for "ID Check" generates a SOAP fault message. Besides "Username" and "Password," we propose two more attributes in the security token for authenticating the user identity in the security token. They are "Subject-Name" and "SubjectLocation," representing the user full name and the postal code of the user's home address. The SOAP fault message describes the exceptional situation of authentication failure because the Web service cannot authenticate the user location based on the "SubjectLocation" in the security token (e.g., 2601) and the "postal-code" in the "IDrecord" (e.g., 2612). Under this circumstance, the integration plan cannot be carried out properly because the "ID Check" is the first critical activity in the workflow.

If the "area-code" is "02" in the user's record and also there is an ontology defining that the area "02" covers the postal codes from 2601 to 2612, the Web service at the "ID Check" activity can authenticate the user location based on the "area-code." This case is referred to the remedy approach in the exceptional handling.

5   Formulation of Workflow View from Flows

Based on the flows identified in the previous section, we can now formulate the workflow view between the interacting organizations. In Fig. 4, we present a workflow view in an investigation of a suspect between intelligence bureau and city bank, which includes the control flow, data flow, semantic flow, security flow and exception flow between these two parties. The XML code is summarized in graphical form with a tool called *XMLSpy* from Altova Inc. (http://www.XMLSpy.com). Because of space limitation, only the control-flow part is depicted in this figure. We proceed to discuss other flows in the following paragraphs.

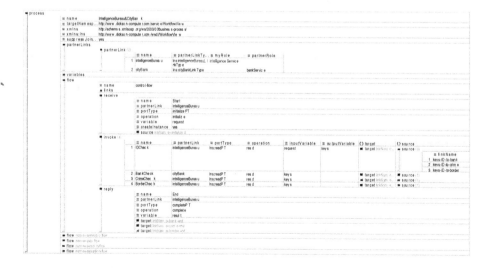

Fig. 4. A Graphical XML Representation of a Workflow View

In the control flow of Fig. 4, the "Start" activity generates a control event "initialized" to trigger the execution of "IDCheck" activity. Similarly, the "IDCheck" activity will trigger the "CrimeCheck" "BorderCheck" and "BankCheck" activities to be executed in parallel with the "keys-transferred" events. In this case, the keys contain a set of records (i.e., "id-no" and "ac-no") for the consequent activities (Web services) to retrieve the dataset. Once these activities are completed, the control-flow is ended successfully.

The data schema can be represented in OWL as an ontology. Fig. 5 shows an OWL ontology in describing the "IDrecord" data schema. As a result, we propose the semantic-referencing assertions in BPEL4WS for supporting semantic-flows as shown in Fig. 5.

```
<owl:Ontology rdf:about="#Identity">

<owl:versionInfo>v 1.00 2003/12/16 22:37:39</owl:versionInfo>
<rdfs:comment>An example OWL ontology for Identity</rdfs:comment>
...
<owl:Class rdf:ID="DataSchema">
<owl:unionOf rdf:parseType="Collection">
  <owl:Class rdf:about="#id-no"/>
  <owl:Class rdf:about="#name"/>
  <owl:Class rdf:about="#sex"/>
  <owl:Class rdf:about="#date-of-birth"/>
  <owl:Class rdf:about="#area-code"/>
  <owl:Class rdf:about="#phone-no"/>
  <owl:Class rdf:about="#address"/>
  <owl:Class rdf:about="#postal-code"/>
  <owl:Class rdf:about="#tax-file-no"/>
</owl:unionOf>
</owl:Class> ...
</owl:Ontology>
```

Fig. 5. A Simplified Data Schema of OWL Ontology

```
<flow name="semantic-flow">
  <ontology activityName="IDCheck">
    <ontologyRef="http://www.example.org/identity.owl" />
  </ontology>
  <ontology activityName="BankCheck">
    <ontologyRef="http://www.example.org/banking.owl" />
  </ontology>
  <ontology activityName="CrimeCheck">
    <ontologyRef="http://www.example.org/legal.owl" />
  </ontology>
  <ontology activityName="BorderCheck">
    <ontologyRef="http://www.example.org/custom.owl" />
  </ontology> ...
</flow>
```

Fig. 6. A Simplified BPEL4WS Code for Illustrating Semantic-Flows

Referring to Fig. 7, the data-flow-1 is used to join (in double arrows) the datasets returned from the "IDCheck," "CrimeCheck," and "BorderCheck" activities into an integrated view for a particular user request. Similarly, the data-flow-2 is used to join the datasets returned from the CrimeCheck," "BorderCheck," and "BankCheck" activities respectively. Using the "id-no" as a join key, the data-flow-1 joins the "IDrecord" dataset (with attributes "id-no," "sex," "age," etc.), the "CrimeRecord" dataset (with attributes "id-no," "Crime-description," "sentence," etc.), and the "BorderRecord" dataset (with attributes "id-no," "entry-or-exit," "place," etc.). Similarly, using the "id-no" as a join key, the data-flow-2 joins the "CrimeRecord," "BorderRecord" and "BankRecord" datasets. In particular, the data linkage (i.e., "id-no" and "tax-file-no") between "BorderRecord" and "BankRecord" are delivered by the "IDCheck" activity from the control-flow. In a general case, the workflow is ended once all the control-flow and data-flow(s) are completed successfully.

```
<flow name="data-flows">                          <integrate name="data-flow-2">
 <integrate name="data-flow-1">                    <dataset name="CrimeRecord">
  <dataset name="IDrecord">                           <attributes name="id-no" key="primary"/>
   <attributes name="id-no" key="primary"/>           <attributes name="crime-description"/>
   <attributes name="sex"/>                            <attributes name="sentence"/>    ...
   <attributes name="age"/>                          </dataset>
                     ...                             <dataset name="BorderRecord">
  </dataset>                                           <attributes name="id-no" key="primary"/>
  <dataset name="CrimeRecord">                         <attributes name="entry-or-exit"/>
   <attributes name="id-no" key="primary"/>           <attributes name="place"/>
   <attributes name="crime-description"/>             <attributes name="date"/>    ...
   <attributes name="sentence"/>                     </dataset>
                     ...                             <dataLinkage name="IDrecord">
  </dataset>                                           <attributes name="id-no" key="foreign"/>
  <dataset name="BorderRecord">                        <attributes name="tax-file-no" key="foriegn"/>
   <attributes name="id-no" key="primary"/>          <dataLinkage/>
   <attributes name="entry-or-exit"/>                <dataset name="BankRecord">
   <attributes name="place"/>                          <attributes name="tax-file-no" key="primary"/>
   <attributes name="date"/>                           <attributes name="bank-no"/>
                     ...                               <attributes name="account-no"/>
  </dataset>                                            <attributes name="transaction"/>    ...
 </integrate>                                        </dataset>
                                                    </integrate>
                                                   </flow>
```

Fig. 7. Proposed BPEL4WS Assertions for Illustrating Data-Flows

A security token in the context of WS-Security is shown in Fig. 8. Based on this, the security flows are defined in Fig. 9 and the exception flows are defined in Fig. 10.

```
<S:Envelope xmlns:S="http://www.w3.org/2001/12/soap-envelope"
       xmlns:wsse=http://schemas.xmlsoap.org/ws/2002/04/secext
       xmlns:wii="http://schemas.workflow.org/wii/2003/12/authentication">
  <S:Header>    ...
    <wsse:Security>
     <wsse:UsernameToken>
       <wsse:Username>93856543</wsse:Username>
       <wsse:Password>3875</wsse:Password>
       <wii:SubjectName>Dickson Chiu</wii:SubjectName>
       <wii:SubjectLocation>2601</wii:SubjectLocation>
      </wsse:UsernameToken>
    </wsse:Security>    ...
  </S:Header>    ...
</S:Envelope>
```

Fig. 8. An Example Security Token

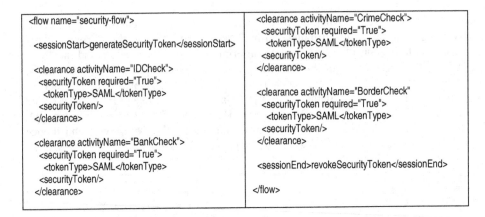

```
<flow name="security-flow">

  <sessionStart>generateSecurityToken</sessionStart>

  <clearance activityName="IDCheck">
    <securityToken required="True">
      <tokenType>SAML</tokenType>
    <securityToken/>
  </clearance>

  <clearance activityName="BankCheck">
    <securityToken required="True">
      <tokenType>SAML</tokenType>
    <securityToken/>
  </clearance>
```

```
<clearance activityName="CrimeCheck">
  <securityToken required="True">
    <tokenType>SAML</tokenType>
  <securityToken/>
</clearance>

<clearance activityName="BorderCheck"
  <securityToken required="True">
    <tokenType>SAML</tokenType>
  <securityToken/>
</clearance>

<sessionEnd>revokeSecurityToken</sessionEnd>

</flow>
```

Fig. 9. A Simplified BPEL4WS Code for Illustrating Security-Flows

```
<flow name="exception-flow">
  <exceptionHandling name="rule-1">
    <event>anyActivitySpecificException</event>
    <condition>affectDataIntegration</condition>
    <action>remedyOrforwardRecoveryProcedure</action>
  </exceptionHandling>
  <exceptionHandling name="rule-2">
    <event>anyCrossActivityException</event>
    <condition>affectDataLinkage</condition>
    <action>backwardRecoveryProcedure</action>
  </exceptionHandling>
  <exceptionHandlingDefault>
    <action>abortControlFlow</action>
  </exceptionHandlingDefault>
</flow>
```

Fig. 10. Proposed BPEL4WS Assertions for Illustrating Exception-Flow

6 Literature Review

There have some earlier works in the area of workflow views and related notions of a partial workflow. Liu and Shen [11] presented an algorithm to construct a process view from a given workflow, but did not discuss its correctness with respect to inter-organizational workflows. Our preliminary approach of workflow views has been presented in [1]. From then, workflow views have been utilized as a beneficial approach to support the interactions of business processes in E-service environment [2, 3]. In [12], we adopted the object deputy model [13] to support the realization of workflow views. Van der Aalst and Kumar [14] present an approach to workflow schema exchange in an XML dialect called XRL but it does not include the support for workflow views. Besides, van der Aalst [15] models inter-organizational work-flows and the inter-organizational communication structures by means of Petri Nets and message sequence charts (MSCs), respectively. Since the author abstracted from data and external triggers, the proposed communication protocol is not as complete as the inter-operation protocol presented in the workflow view approach [3]. To address the derivation of private workflows from inter-organizational workflows, Van der Aalst and Weske [16] uses the concept of workflow projection inheritance introduced in [17]. A couple of derivation rules are proposed so that a derived workflow is behaviorally bi-similar to the original workflow based on branching semantics, in contrast to the trace semantics adopted in the workflow view model. Schulz and Orlowska [18] propose to tightly couple private workflow and workflow view with state dependencies, whilst to loosely couple workflow views with control flow dependencies. They also develop a cross-organizational workflow architecture for view-based cross-organizational workflow execution.

Prior research has proposed abstracting the information integration problem into querying an infrastructure mediation service [19], which offers users and applications a location-independent virtual integrated schema in a common data model [20]. Although information integration issues are not new in database research communities, applying workflow technologies in different application domains has many unique properties that entail special integration design considerations, such as [21]. Cheung et al. [22] use a bottom-up data-driven methodology to extend information systems into Web services. However, this paper presents a top-down approach and focus on a global view of the process.

Recently, the Business Process Execution Language for Web Services (BPEL4WS) [4], a formal specification of business processes and interaction protocols, has been proposed. BPEL4WS defines an interoperable integration model that facilitates the expansion of automated process integration in both intra- and intercorporate environment. In particular, the current version of BPEL4WS claims that data flow will be allowed through links in addition to using links to express synchronization dependencies in the future version [4]. Therefore, we demonstrate the proposed models with our proposed data-integration, semantic-referencing, and exception-handling assertions in the context of BPEL4WS. In summary, all these XML languages facilitate defining Web services interacted activities in the format of a workflow. However, except SOAP-fault captures exceptions in the message level. Further, these languages do not provide any expression to capture exceptions comprehensively.

Exception issues have been widely investigated in the workflow research community. For example, Hwang et al. [23] propose a model for handling workflow exceptions. The proposed model provides a rule base that consists of a set of rules for handling exceptions. If none of the rules match the current exception, a search on the previous experience in handling similar exceptions is conducted. They also describe several algorithms to identify the exception records by classifying the kind of information about exceptions, defining the degree of similarity between two exceptions, and searching similar exceptions. Similarly, Casati and Pozzi [24] present a methodology for modeling exceptions by means of activity graphs. They describe taxonomy of expected exceptions by categorizing and mapping them into activity graphs. They also show how to handle the exceptions in each class. Further, they also provide methodological guidelines in order to support exception analysis and design activities. Based on a taxonomy and meta-model, Chiu et al. [25,26] developed a web-based WFMS, called ADOME-WFMS, to support automatic resolution for expected exceptions and human intervention for unexpected exceptions, through a unified framework of event-condition-action (ECA) rules. Advanced matchmaking was also supported with a role and capability model. However, all of these works do not explicitly separate exception-flows from the control-flows and data-flows.

The Semantic Web is originally based on the research areas of knowledge representation and ontology in Artificial Intelligent (AI). DAML+OIL is a semantic markup language based on RDF and RDF Schema extended with richer modeling primitives [27]. DAML+OIL provides a language for expressing far more sophisticated classifications and properties of resources than RDFS [28]. Very recently, the OWL Web Ontology Language is being developed by the W3C Web Ontology Working Group as a revision of the DAML+OIL web ontology language. OWL [29] has been proposed to provide three increasingly expressive sub-languages for specific communities of implementers and users, namely, OWL Lite, OWL Description Logics (OWL DL), and OWL full. OWL Lite supports the basic need for a classification hierarchy and simple constraints. For example, while it supports cardinality constraints, it only permits cardinality values of 0 or 1. Thus, OWL Lite provides an easier implementation and a quicker migration path for thesauri and other taxonomies. OWL DL supports maximum expressiveness while retaining computational completeness (all conclusions are guaranteed to be computed) and decidability (all computations will finish in finite time). OWL DL includes all OWL language constructs, but they can be used only under certain restrictions (for example, while a class may be a subclass of many classes, a class cannot be an instance of another class). OWL DL is so named due to its correspondence with description logics, a field of research that has studied the logics that form the formal foundation of OWL. OWL Full supports maximum expressiveness and the syntactic freedom of RDF, but has no computational guarantees. For example, in OWL Full a class can be treated simultaneously as a collection of individuals and as an individual in its own right. OWL Full allows an ontology to augment the meaning of the pre-defined (RDF or OWL) vocabulary. Thus, ontology developers adopting OWL should consider which sub-language best suits their needs. In this paper, ontology is described in OWL, in particular in OWL DL, because OWL provides a standard set of elements and attributes with defined semantics, for defining terms and relationships in ontology. In addition, OWL contains a set of logic-based primitives that are specifically useful in intelligence informatics. Furthermore, we decided to deploy OWL instead of DAML+OIL because OWL has been designed as a standard in W3C [30].

In summary, the development of solutions for workflow-based information integration is promising and challenging. To our knowledge, none of the prior research studies the use of workflow technologies to materialize information integration from the control-flows, data-flows, security-flows, and exception-flows in a unified approach. Such application in security and intelligence informatics is novel. Neither have there been discussions on the methodology for systematic partial or restricted workflow formulation.

7 Conclusions

This paper has proposed a new perspective of workflow views through a subset of various flows of original workflow. As such, in additional to basic control flow, workflow views are now enriched with the support of data flow, semantics flow, exception flow, and security flow. We believe that this is a viable solution for systematic design of workflow views for better B2B interaction specification and management. This application of the "separation of concerns" principle enables better understanding of application semantics and is especially useful for large-scale information systems. Our approach is still extensible because the methodology is still valid even when new types of flows (say, privacy flow) are identified.

We have also detailed each type of flows, their usage, and possible implementation with contemporary Web services technologies. Further, we identified some useful relationships among flows. In particular, we have demonstrated how semantics flows can help exception handling. An in-depth study of this topic is of paramount interest as this should be one of the most useful applications of Semantic Web technologies to workflows.

There are more issues that can be explored to expand this work. In particular, privacy-flow relationships describe each service's data practices what information they collect from individuals and what (e.g., purposes) they do with it. Please note that this paper takes privacy-flow as future work. In particular, security-flows and privacy-flows are conflicting but both required according to laws and regulations. Further, we are working on alerts, that is, process urgency requirements, which is also in line with the flow concept. We are also interested in further studies in requirements engineering aspects of our approach.

References

1. Chiu, D. K. W., Karlapalem, K., Li, Q.: Views for Inter-organization Work-flow in an E-commerce Environment. In: Proc. Semantic Issues in E-Commerce Systems, IFIP TC2/WG2.6 Ninth Working Conference on Data-base Semantics (2001)
2. Chiu, D. K. W., Karlapalem, K., Li, Q., Kafeza, E.: Workflow View Based E-Contracts in a Cross-Organizational E-Services Environment. Distributed and Parallel Databases 12 (2002) 193-216
3. Chiu, D. K. W., Cheung, S. C., Till, S., Karlapalem, K., Li, Q., Kafeza, E.: Workflow View Driven Cross-Organizational Interoperability in a Web Service Environment. Information Technology and Management to appear (2004)
4. BPEL4WS. Available: http://www.ibm.com/developerworks/webservices/library/ws-bpel/
5. OWL. Available: http://www.w3c.org/2004/OWL/

6. Leymann, F., Roller, D.: Using Flows in Information Integration. IBM Systems Journal 41 (2002) 732-742

7. WS-Security. Available: http://www.oasis-open.org/committees/wss

8. SAML. Available: http://www.oasis-open.org/committees/security

9. Chiu, D. K. W., Li, Q., Karlapalem, K.: Facilitating Exception Handing with Recovery Techniques in ADOME Workflow Management System. Journal of Applied Systems Studies 1 (2000) 467-488

10. Perry, D. E., Romanovsky, A., Tripathi, A.: Current trends in exception handling. Software Engineering, IEEE Transactions on 26 (2000) 921-922

11. Liu, D.-R., Shen, M.: Modeling workflows with a process-view approach. In: Proc. Database Systems for Advanced Applications, 2001. Seventh International Conference on (2001) 260-267

12. Shan, Z., Long, Z., Luo, Y., Peng, Z.: Object-oriented Realization of Workflow Views for Web Services - an Object Deputy Model Based Approach. In: The Fifth International Conference on Web Age Information Management, WAIM 2004.

13. Kambayashi, Y., Peng, Z.: An Object Deputy Model for Realization of Flexible and Powerful Object-bases. Journal of Systems Integration 6 (1996) 329-362

14. Aalst, W. M. P. v. d., Kumar, A.: XML Based Schema Definition for Support of Interorganizational Workflow. Information Systems Research 14 (2003) 23-46

15. Aalst, W. M. P. v. d.: Inter-organizational Workflows: An Approach based on Message Sequence Charts and Petri Nets. Systems Analysis - Modeling - Simulation 34 (1999) 335-367

16. Aalst, W. M. P. v. d., Weske, M.: The P2P Approach to Inter-organizational Work-flows. In: Proc. 13th International Conference Advanced Information Systems Engineering (CAiSE 2001) (2001) 140-156

17. Basten, T., Aalst, W. M. P. v. d.: Inheritance of Behavior. Journal of Logic and Algebraic Programming 47 (2001) 47-145

18. Schulz, K. A., Orlowska, M. E.: Facilitating cross-organizational workflows with a workflow view approach. Data & Knowledge Engineering 51 (2004) 109-147

19. Wiederhold G.: Mediators in the architecture of future information systems. IEEE Computer, vol. 25, no. 3, pp. 38-49, March 1992.

20. Sheth A., Larson J.: Federated database systems. ACM Computing Surveys, vol. 22, no. 3, pp. 183-236, 1990.

21. Sheng O. R. L., Chen G. H. M.: Information management in hospitals: An integrating approach. Proceedings of Annual Phoenix Conference, pp. 296-303, 1990.

22. Cheung S.C., Chiu D.K.W., Till S.: A Data-driven Methodology to Extending Workflows across Organizations over the Internet. Proceedings of 36th Hawaii International Conference on System Sciences (HICSS36), CDROM, 10 pages, IEEE Computer Society Press, Jan 2003.

23. Hwang, S. Y., Ho, S. F., Tang, J.: Mining exception instances to facilitate workflow exception handling. Proceedings of the 6th International Conference on Database Systems for Advanced Applications, pp. 45-52, 1999

24. Casati, F., Pozzi, G.: Modeling Exceptional Behaviors in Workflow management Systems, Proceedings of International Conference on Cooperative Information Systems (CoopIS'99), 1999

25. Chiu, D. K. W., Li, Q., Karlapalem, K.: A Meta Modeling Approach for Workflow Management System Supporting Exception Handling. Information Systems, vol. 24, no. 2, pp. 159-184, May 1999

26. Chiu, D. K. W., Li, Q., Karlapalem, K.: Web Interface-Driven Cooperative Exception Handling in ADOME Workflow Management System. Information Systems, vol. 26, no. 2, pp. 93-120, 2001

27. RDF: An Axiomatic Semantics for RDF, RDF-S, and DAML+OIL. March 2001. Online: www.w3.org/TR/daml+oil-axioms

28. DAML: The DARPA Agent Markup Language (DAML) Program. 2003. Online: http://www.daml.org

29. Web-Ontology (WebOnt) Working Group: www.w3.org/2001/sw/WebOnt

30. Chen, H., Finin T., Joshi, A.: Using OWL in a Pervasive Computing Broker. Available Online: citeseer.nj.nec.com/583175.html

A Practice in Facilitating Service-Oriented Inter-enterprise Application Integration

Bixin Liu, Yan Jia, Bin Zhou, and Yufeng Wang

National University of Defense Technology, Changsha, China
{bxliu, binzhou, yanjia, yfwang}@nudt.edu.cn

Abstract. Web services have been enjoying great popularities in recent years. This paper addresses the issue of introducing web services to conventional middleware-based enterprise applications to facilitate inter-enterprise integration. StarWebService is presented as a framework to support service-oriented integration by gracefully bridging the gap between web services and middleware. On the basis of a hierarchy resource model and runtime infrastructure, StarWebService enables not only exporting enterprise resources as web services but also importing web services as enterprise resources. It offers an economical and flexible way to provide and consume services for cross-organizational integration.

1 Introduction

Middleware has been booming in the past ten years as the solution for heterogeneous system integration within enterprises. Lots of enterprise information systems have been built based on middleware infrastructures such as CORBA and J2EE. With the increasing requirement for cross-organizational collaboration, web services [1] and service-oriented computing [2,3] have been proposed as the mainstreaming paradigm for loose-coupled integration beyond the enterprise boundaries. This trend demands enterprises to accommodate their conventional middleware-based systems to service-oriented architecture so as to facilitate comprehensive integration with their business partners.

Although this issue has been addressed by some famous enterprise integration vendors (like BEA, IBM, IONA, Microsoft) by upgrading their middleware products with the ability to create web services [5,6,7,8,9] from existing applications, problems are only partly solved. We separate the integration requirements into two aspects. On one hand, a general and incremental approach for exposing internal middleware-based systems as web services is crucial for saving IT investment and migrating application infrastructure to web services technology quickly and economically. On the other hand, enterprises are also likely to outsource services from their business partners. Consistent integration of the imported services with the middleware-based infrastructure of inner systems is desirable for minimizing modifications to existing systems and decreasing risks of system evolution. In short, integration across enterprise boundaries demands introducing web services gracefully into existing infrastructures, so as to benefit both from web services and mature middleware technologies and legacy systems.

M.-C. Shan et al. (Eds.): TES 2004, LNCS 3324, pp. 122–135, 2005.

In this paper, we present our ongoing work on service-oriented enterprise application integration framework with the prototype system StarWebService. It is designed based on a layered resource view and runtime infrastructure. Characterized by its novel service runtime environment and bidirectional resource gateways, StarWebService closes the gap between web services and underlying specific middleware technologies and enables rapid inter-enterprise integration without additional coding. It compliments works from traditional middleware domain and web service domain, and facilitates integration across organizations.

The paper is organized as follows. In section2 we present the main idea and system overview of StarWebService. Then in section3 and section 4, we respectively introduce in detail the service runtime environment with bus-container-service architecture and bidirectional resource gateways which bridge web services with various middleware technologies. With an example of procurement, section 5 demonstrates bidirectional inter-enterprise integration with StarWebService. After introducing related work in section6, we draw conclusions in section 7 with a summary and discussion on future work.

2 Overview of StarWebService System

Our research group [21] has been concentrating on enterprise computing and middleware technologies for several years. StarWebService, a new member in the Star family, targets at providing approaches for bringing the conventional middleware technologies and web services together.

2.1 General Ideas and Main Concepts

StarWebService project was started with the belief of symbiosis of web service and conventional middleware technologies. Although web service is regarded as the most promising infrastructure for next generation of heterogeneous resources integration, the process of adopting this new technology is always evolutional. As a result, the needs for migrating existing middleware-based enterprise applications to service-oriented architecture gracefully become obvious and urgent. A general platform and corresponding tools will be of great help to facilitate such migration, which forms the original motivation of the project.

Before the system details are described, we'd like to illustrate the general ideas and main concepts in StarWebService.

StarWebService is designed on the basis of a layered resource view, which reflects how enterprise applications are established and organized. In conventional middleware-based applications, application components are inter-connected according to specific middleware infrastructure. We call those components conforming to specific middleware technologies *enterprise resources*. For instance, CORBA components, EJBs are all categorized as enterprise resources. Enterprise resources can be established from scratch or by wrapping various fabric resources and legacy applications. Thanks to middleware, relative homogeneity is achieved within enterprise scale. However, when integration across enterprise boundaries is considered, heterogeneity in middleware infrastructures of autonomous organizations brings new challenges. So web services are introduced at the enterprise boundaries to close the gaps. We model the resources shared across organizations as *enterprise services*, which are technology and platform independent function modules with well-defined interfaces. Enterprise services simplify integration by shielding partners from technology details of inner-enterprise infrastructures and establish a common environment for interactions. Finally, at the top of the resource layer are *business processes*, constructed by composing enterprise services, which provide composite value-added services. But business process is not emphasized in this paper. Further discussions can be found in [20].

The belief of symbiosis of web service and middleware is fulfilled in StarWebService by enabling bidirectional mapping between enterprise resources and enterprise services.

On one hand, when enterprise resources are to be shared across organizations, StarWebService helps to construct enterprise services as the external views for them. Standard service descriptions are generated from original formats of enterprise resource descriptions. Those services are in some sense virtual because they employ existing enterprise resources as their implementations. External requests to services are transparently translated to technology-specific formats and then dispatched to the right backend resources. In this way, enterprise resources are exported to business partners as enterprise services which utilize enterprise resources to perform their functions.

On the other hand, there are cases that external services are expected to be invoked by inner system as technology-specific resources. A typical scenario is that a service from a business partner is outsourced as substitute of an out-of-date component within enterprise system. It may become a tragedy if the discarded module is tangled with other parts of the systems and all of them have to be modified in order to interact with the service. The capability of importing services as specific enterprise resource will simplify the problem in this situation and help to maintain consistence during system evolution. Similarly but reversely, StarWebService creates virtual enterprise resources that act as internal views for the imported services and delegate all the invocations on those resources to corresponding services.

2.2 StarWebService System

Fig.1 shows the layered runtime infrastructure of StarWebService system. The bottom depicts conventional middleware-based applications within enterprises. Application

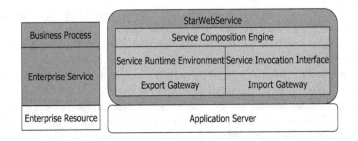

Fig. 1. StarWebService architecture

server plays the role of runtime environment for enterprise resources. There are many application server products from different vendors. Although the application server does not belong to StarWebService, it is included for completeness.

The area filled with dark grey in Fig.1 shows the case across organizational boundaries, which is emphasized by StarWebService. Compared to the application server for inner-enterprise applications, *service runtime environment (SRE)* is provided as the server side platform for deploying, running and managing enterprise services. It is the kernel of StarWebService, consisting of components for protocol processing, dynamic service deploying, service management and monitoring and etc. Because we consider services built incrementally from enterprise resources, the SRE is specially designed to provide service façades for enterprise resources. As a result, the process of service development is simultaneous with service deployment. We will show details of SRE in the next section. As the counterpart of SRE, *service invocation interface (SII)* provides a group of APIs for client applications to invoke services with SOAP messages.

To facilitate transparent bidirectional mapping across resource layers, the resource gateways are introduced into StarWebService system. They deployed at the edge of enterprise systems serving for connecting specific middleware technology with web services. Their working directions differentiate them as *export gateways* for exposing enterprise resources as services and *import gateways* for outsourcing services as enterprise resources. The main functions of resource gateways include converting description between enterprise services and specific enterprise resource type, translating message from SOAP to specific interoperation protocol (like IIOP) or vice versa, and adapting invocation to backend enterprise resources or services. The export gateways are registered to the SRE and invoked when enterprise resources are deployed as services and when those services are invoked, while the import gateways utilize the SII to invoke external service. Section 4 will show the details of bidirectional resource gateways.

At the top of the runtime infrastructure is the *service composition engine*, which composes enterprise services into business processes. We argue that a non-centric service composite execution is critical for scalability. So a collaborative process execution mechanism is adopted in the composition engine, presented in [20]. It's beyond the scope of this paper.

In the following sections we will present the details of service runtime environment and bidirectional resource gateways, the most interesting parts in StarWebService.

3 Service Runtime Environment

The service runtime environment denotes a software layer that offers fundamental supports for deploying, running and managing enterprise services. It can be compared to a special application server that supports web services.

Container-component model [17, 18] has been proved a successful architecture for constructing CORBA and J2EE application servers, which inspires us to apply similar concept to design SRE. But SRE have to deal with more complicated situations, for example, supporting multiple underlying transport protocols for SOAP messages and connecting to various kinds of back-end resources that implement services. So we propose the bus-container-service architecture for SRE, as shown in Figure2.

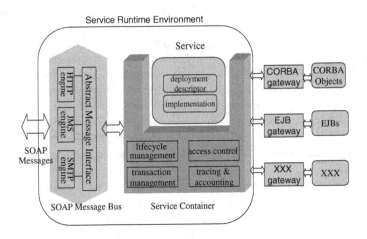

Fig. 2. Service runtime environment architecture

3.1 SOAP Message Bus

SOAP message bus is the communication infrastructure of SRE, enabling transport independent SOAP message exchange. It is designed for shielding other components in SRE from details of multiple transport bindings and providing consistent interfaces for sending, receiving, packaging and extracting SOAP message.

It is composed of an *abstract message interface* and multiple *transport engines*. The abstract message interface provides a uniform representation for SOAP messages and simple, transport-agnostic interfaces for manipulating them. However, a transport engine serves for specific underlying transport protocol. They listen on specific ports, parse specific protocol packages, extract SOAP envelopes, hand them over to the abstract message interface, and vice versa. A transport engine can simply consist of a listening thread and several working threads, or be a complicated one with complex connection management mechanism to improve their scalability and performance. An important approach is to implement transport engines based on some components of existing advanced servers. For example, web container supported by most web servers is a good choice for implementing the SOAP/HTTP engine.

3.2 Service Container

Service container is the hosting environment for service instances. It isolates service implementation from the services runtime infrastructure, and provides fundamental support service instance creation, running and management. Primary functions of service container include:

- **Service instance lifecycle management.** When a request arrives, service container creates an instance and management its lifecycle according to life scope declaration in the service's deployment descriptor. Three kinds of life scope are supported: request, the instance is created when request arrives and destroyed after response returns; session, the instance works for a whole session and the service container is responsible for maintaining the conversation status; application, once the instance is created it won't be destroyed until the server process ends.
- **Resource gateway registry.** Service container manages available resource gateways, and provides call back interface for the service instances to find and invoke proper gateway.
- **Service access control.** To insure secure access to sensitive services, role-based access control is imposed before requests are processed by services. Service container maintains a policy set which is used to decide the validity of each request. Both service level and operation level control are supported.
- **Transaction management.** Service container also takes the responsibility of propagating transaction context and coordinating transactional resources to commit or rollback.
- **Tracing and accounting.** Because all messages are delivered to service container before processed by the service, it is proper to do tracing and accounting in the container. Further management can be carried out with collected information.

Function modules of service container are designed as pluggable components with extensible interfaces, which enable on-demand system configuration and extension by plugging new modules.

Note that a service container is dedicated to a specific transport engine in the SOAP message bus, and responsible for processing incoming requests from that engine. So after a service container is installed, it has a unique transport-specific location. A transport engine may have more than one containers associated with it, and they are distinguished by their names or paths.

3.3 Services

A service is conceptually a "component" deployed in the service container with its deployment descriptor and implementation.

The deployment descriptor of a service, similar with that of EJB, records its meta information, including its identifier, lifecycle scope, resource type, location of its implementation, and some other properties. However in StarWebService, the implementation of a service refers to enterprise resources that actually perform functions declared by the service. They can be collocated with service container, but commonly they can be any kind of applications running anywhere. Implementation is logically related to the service by specific export resource gateway, which delivers all the incoming invocations to corresponding back-end resource. We will go to the details in the next section.

Similarly, a service container can host more than one service. From the consumers' view, those services run at the same location but have different identifiers.

4 Bidirectional Resource Gateways

As we have mentioned in section 2, StarWebService bridges various middleware infrastructures with web services by constructing virtual views between specific enterprise resource type and enterprise service. The export and import gateways in the system offer bidirectional mapping across resource layers which differs from works in [10, 11, 12] which only concern with one-way adaptation.

Generally speaking, a resources gateway performs two functions. One is to convert the descriptions between certain enterprise resource type and enterprise service. The other is to adapt invocation to backend enterprise resources (services) in proper protocols.

Up to now, we have implemented export gateways for typical enterprise resource types including EJB, CORBA objects, CCM [18] and JMS client as well as import gateway for CORBA resource. In the remainder part of this section, we will take CORBA resources as the example to observe the details of resource gateways.

4.1 Gateway Components

The export and import gateways for CORBA resources consist of counterpart groups of components, as Fig.3 shows:

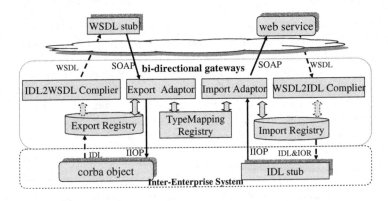

Fig. 3. Bidirectional resource gateways for CORBA applications

- *Export/import registries* maintain logical associations between services and CORBA objects. Typically, pairs of endpoint mappings are stored in the registries to resolve CORBA objects that implement services or services that are masked by virtual CORBA objects.
- *Type mapping registry* deals with data conversion between IDL and XML data types. It provides standard schemas and custom type mappings that enable compliers and adaptors to convert declared data type and encoded data stream successfully.
- *IDL2WSDL/WSDL2IDL compilers* generate WSDL documents from IDL files or vice versa. The compliers access the export or import registry during the conversion for necessary information, such as the identity of the exposed service.

Type mapping registries are also called to convert data type declarations when compiling.

- *Export/import adaptors* are the core components of the gateways. They serve as the mediators for interoperation across IIOP and SOAP protocol domains by adapting invocations to the right targets with the right encoding format. Only RPC style is considered here, so adaptors translate request/response messages and exceptions messages.

4.2 Export CORBA Objects as Services

We explain the mechanism for exporting CORBA objects as two phrases, shown in the left part of Fig.3.

The first is deploying phrase, depicted with dashed arrows. During this phase, the CORBA object to be exposed is registered to the export gateway with its IOR and IDL file. Moreover, a global unique qualified service name is provided. The registration will trigger actions in the export registry and IDL2WSDL compiler. For the export registry, mapping of <IOR, service_name> is constructed. Besides, interface information in IDL files is parsed and saved to a repository in the registry (implemented as the Interface Repository Service in CORBA), which is then utilized by the compiler to generate WSDL document. After that the exposed service is deployed to the service container in charge of the export gateway with a record in the deployment descriptor <service_name, resource_type >. Now the first phrase ends.

Several points need to be mentioned roughly. For compilation, conversion rules regarding all the IDL syntax are compliant to standard from OMG [19]. Constructed data types in IDL are treated as custom defined XML types registered to the type mapping registry which ensures they can be recognized and converted successfully when the parameters of those types are received in the execution phrase. And in the result WSDL description, the location of the exported service actually refers to the address where the charging SRE runs.

The execution phrase, depicted with solid arrows in Fig.3, begins when the consumer of the exported service generates stubs from the published WSDL and makes invocation with SII. Because the generated WSDL is standard, other web service platforms such as .NET can be used to develop client applications too. Fortunately SOAP requests will firstly be directed to the SRE indicated by the service location and then to the export gateway. Now the following tasks are performed at the export adaptor to translate a SOAP request to an IIOP request: get the IOR of the target CORBA object according to the entries in the export registry (locate); map the SOAP-encoded parameters to CORBA compliant variables by the aid of type mapping registry (decode); construct and emit IIOP request (invoke). When the IIOP response (or exception) is returned, it is translated back to a SOAP response (or fault) in a similar way.

Note that because the mappings between SOAP types and IDL types are not 1 to 1, additional information from the repository in the export registry will be used to determine the target IDL types of given SOAP parameter at the execution phrase.

4.3 Import Services as CORBA Resources

The mechanism of importing services resembles that of exporting resources, except that the client needs both IOR and IDL to invoke the virtual CORBA object for the imported service. So not only IDL is generated from WSDL when deploying a service to the import gateway, but also the IOR for the virtual object is created, as the right of Fig.3 shows.

The deploying phrase starts with compiling the WSDL document of the imported service to an IDL file. The import registry records the pair of <service_name, service_location> after parses the WSDL. At the same time, the generated IDL information is saved to a repository in the import gateway which is later necessary for the virtual CORBA objects to marshal and unmarshal data in IIOP messages.

To achieve dynamic creation of virtual CORBA objects, we design a special servant registered to a POA with policy of USER_OBJECT_ID in the import gateway. It creates a new mapping of <service_name, object_id> in the import registry instead of creating an object instance when a new service is registered to the gateway. The object_id is generated automatically and the gateway ensures its uniqueness. It is then used to make IOR for the virtual object for the imported service.

In the execution phrase, with the generated IDL and IOR, the CORBA client can make invocation in the same way as invoking an ordinary CORBA object, without any knowledge about web services or SOAP. Evidently, the IIOP requests to the virtual object will be delivered to the import gateway, where they are translated and adapted to the external service in SOAP.

5 Inter-enterprise Integration with StarWebService

In this section we will take a simple example of procurement to demonstrate how flexible and economical inter-enterprise integration is achieved by StarWebService.

In this case, shown in Fig.4, information system in enterprise ABC, employing CORBA technology, has been operating for several years. It is an integrated system that combines applications in the sale, finance, human resource, shipping and other

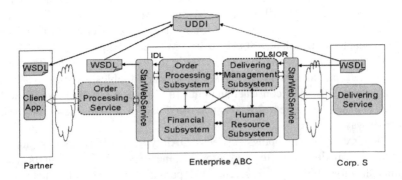

Fig. 4. The procurement example

departments. With its business development, the enterprise wishes to expose the order processing subsystem as web services to enable high level supply chain integration. Meanwhile, delivering service from corp. S is outsourced as the result of organization reform that cut the shipping department to improve efficiency. We'll show how StarWebService works in this scenario with minimum influence on existing systems.

Firstly, CORBA objects in the order processing subsystem that needs to be exposed as services are decided. Because all objects in this subsystem do not deal with placing orders, only those directly interacting with buyers are identified. Then the operations of selected objects are examined to ensure that they are proper to be invoked by business partners. After the set of objects and their operations are determined, their IDLs are collected as an input IDL file to start StarWebService. As described in the previous section, StarWebService registers an order processing service in its service container, compiles the IDL file and generates WSDL document which can be published to the UDDI. Later when the partner invokes the order processing service, requests are translated into IIOP messages by StarWebService and adapted to corresponding objects in the order processing subsystem.

StarWebService provides wizard for accomplish works described above. Fig.5 shows a snapshot of the wizard for exporting CORBA resource as a service. All the developers need do is to fill the deployment form with correct information. Because of dynamic deploying and adapting mechanism supported in StarWebService, zero-coded integration is achieved.

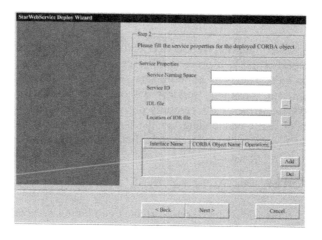

Fig. 5. Export wizard in StarWebService

To import delivering service from Corp.S, its WSDL description is firstly retrieved (e.g. from UDDI). Then it is passed to StarWebService to create a virtual delivering object with its IOR and IDL definitions. Now that the delivering object interacts with other subsystem substituting for the external service in Corp.S, there won't be dramatic modifications in other subsystems, especially when mechanism like naming service are used in the enterprise system.

Similarly a graphic studio is included in StarWebService system to guide the developers to manage imported services. It integrates the import gateway and other tools to support the whole loop of deploying imported services, developing client programs, debugging and running the whole application. Fig.6 shows part of its interface.

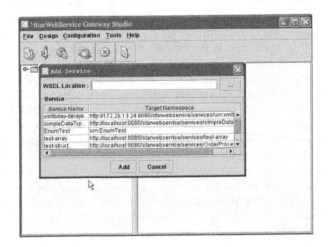

Fig. 6. Studio for importing web services as CORBA objects

This example shows that StarWebService facilitates inter-enterprise integration in the following aspects: firstly, it protects IT assets in enterprises against technology revolution to the most extend by building services incrementally from middleware-based application and consuming services without great impact on existing inner systems; secondly, in virtue of it's carefully designed platform and tools, it enables integration without additional coding, so the complexity of dealing with various middleware technologies can be reduced and efficiency can be improved.

6 Related Works

There are many works addressing service-oriented computing in recently years from both industries and research groups. Also practices have been done to apply web services to inner-enterprise and inter-enterprise applications [4]. Famous enterprise integration vendors like Microsoft, IBM, BEA and SUN have extended their product lines to support this new technology. StarWebService is similar with them in the sense of a platform for developing and running web services. However, few of those commercial products have addressed the migration issues, especially from bidirectional aspects. The relationship between conventional middleware and web services is still an open issue. We make a point that web services and middleware technologies, like CORBA and J2EE, will coexist and benefit each other. For this reason we present a framework to accommodate middleware-based applications to web services for inter-enterprise integration.

Most commercial products on web services are tightly bound to specific platform like .NET or J2EE. A few works also emphasis on the general framework for bringing conventional enterprise systems and web services together. Typically, CapeClear [11] and open source project Axis [12] resemble StarWebService in building web service from several types of enterprise resources. However, StarWebService provides more comprehensive support for various enterprise resource types than they do.

Discussions on the architecture and techniques for building service runtime environment are not as plentiful as expected. The bus-container-service architecture presented in this paper is justified to be a feasible way. Although the concept of service container is also addressed in OGSA [16], the container's primary responsibility is to ensure the services adhering to Grid service semantics. We propose service container for providing common service semantics that satisfy the requirements of enterprise computing, hence properties like security and transaction managements are concerned.

There are also some works that approach to web service for other purposes. [13] introduces a framework for on-the-flying wrapping of Jini resources as web services, [14,15] is presented for the similar purpose. While we emphasize enterprise resources which is based on popular enterprise middleware infrastructures.

The last but not the least, works mentioned above only concentrates on wrapping enterprise resource as services, neglecting the requirement of outsourcing services consistently. Considerations for the import gateways are absent from their works. StarWebService exceeds them in that it addresses bidirectional integration so that it combines web services and middleware technologies integrally.

7 Conclusions

Popularity of service-oriented computing calls for migrating existing enterprise applications to this new paradigm. In this paper, we first address the issue of introducing web services to conventional middleware-based enterprise applications, and then present our work on integrating web services and conventional middleware technologies in order to facilitate inter-enterprise integration.. By virtue of StarWebService, conventional middleware based applications within enterprises can be easily and quickly delivered to the business partners as web services. Also web services from business partners can be utilized by internal applications without changing existing system infrastructure. It can be of great help to protect investment and reduce difficulties in developing inter-enterprise applications. As complement for works from middleware domain and web service domain, StarWebService provides a valuable solution for flexible and economical inter-enterprise integration.

The future work will mainly focus on two aspects. The first is to improve the system with monitoring mechanism across vertical layers and later a management framework for enterprise services and enterprise resources. The other is to apply StarWebService to more domains. Technologies for automatic service discovery and composition are also in our research plans.

Acknowledgements

This work is supported by the National Natural Science Foundation of China under Grant No.90104020, the National High-Tech Research and Development Plan of China under Grant No.2002AA116040 and the National Grand Fundamental Research 973 Program of China under Grant No.G1999032703.

References

1. David Booth, Michael Champion, etc. "Web Services Architecture" W3C Working Draft, May 2003. http://www.w3.org/TR/2003/WD-ws-arch-20030514/
2. M.P. Papazoglou and D. Georgakopoulos, "SERVICE -ORIENTED COMPUTING", COMMUNICATIONS OF THE ACM, October 2003/Vol. 46, No. 10 pp.25-28
3. Mike P. Papazoglou. "Service-oriented computing: Concepts, characteristics and directions". In Proc. of 4th "International Conference on Web Information Systems Engineering (WISE 2003), DEC 10-12, 2003, pp.3-12.
4. Steve Vinoski, "Integration with Web Services", IEEE Internet Computing, Nov-Dec 2003 issue.
5. BEA Corporation, WebLogic Workshop 8.1 documents, http://edocs.bea.com/workshop/ docs81/index.html , 2004
6. IBM, WebSphere web services , http://www-106.ibm.com/developerworks/websphere/ zones/webservices
7. IONA Corporation. "Service Oriented Integration: A Strategy Brief'. White paper, Jan 2004. IONA Corporation, Artix Enterprise Web Service, http://www. iona.com/products/ artix/artix_prod_enterprise_web_services.htm
8. Microsoft, .NET Framework, http://msdn.microsoft.com/webservices/downloads/ default.aspx
9. Sun, Web Services Developer Pack (WSDP), http://java.sun. com/webservices/downloads/ webservicespack.html
10. Chandra Venkatapathy□Simon Holdsworth. An introduction to Web Services Gateway, http://www-106.ibm.com/developerworks/websphere/zones/ webservices, 2002
11. "Web Service-Oriented Archetecture: The Best Solution to Business Integration", White paper of Cape Clear Software, http://www.capeclear.com
12. Apache Software Foundation, Axis project, http://ws.apache. org/axis/
13. Gannod GC, Zhu HM, Mudiam SV. "On-the-fly wrapping of Web Services to support dynamic integration". Proceedings of 10th Working Conference on Reverse Engineering (WCRE 2003), NOV 13-16, 2003, pp.175-184
14. Yan Huang and David W. Walker. "Extensions to Web Service Techniques for Integrating Jini into a Service-Oriented Architecture for the Grid". In Computational Science - ICCS 2003 (Part 3),published by Springer Verlag as Lecture Notes on Computer Science, vol. 2659, pages 254-263, 2003. ISBN 3-540-40196-2.
15. Y.Huang. "JISGA: A Jini-based Service-oriented Grid Architecture," The International Journal of High Performance Computing Applications, vol. 17, no. 3, 2003. pp. 317-327. ISSN 1094-3420.
16. I. Foster, C. Kesselman, J. M. Nick, and S. Tuecke. "Grid Services for Distributed System Integration". IEEE Computer, 35(6), 2002.
17. Sun Microsystems Inc. "Java™ 2 Platform Enterprise JavaBeans™ Specification, v2.1", Final Draft ,2002

18. Object Mnagement Group, CORBA Component Model Specification. 2002.6
19. CORBA to WSDL/SOAP Interworking   Specification http://www.omg.org 2003.1
20. Bixin Liu, YuFeng Wang, etc. "Collaborative Process Execution for Service Composition with StarWebService", in Proc. of NPC2004, LNCS, Springer, to be published.
21. StarMiddleware Group, StarWebService project, http://www.starmiddleware.net

Discovering and Using Web Services in M-Commerce

Debopam Acharya, Nitin Prabhu, and Vijay Kumar

SCE, Computer Networking,
University of Missouri-Kansas City,
Kansas City, MO 64110
dargc(npp21c, kumarv)@umkc.edu

Abstract. Web Services is slowly evolving to be a promising technology for developing application in open, loosely coupled and distributed computing environments and in mobile commerce. The web is no longer a repository of information and has evolved into a medium for providing general and user-specific services to customers. One of the key requirements of the web services to meet user expectation is its universal accessibility free from temporal and spatial constraints. Such accessibility is not easy to achieve through wired internet and it appears that mobile approach is the only way out. Thus to access the web services from anywhere and anytime, a suitable wireless web services architecture is needed which must overcome the limitations of the mobile environment (service discovery, low bandwidth, limited power source and scalability bottleneck). This paper exploits the advances in wireless broadcast discipline and proposes a new architecture that overcomes a number of problems in discovering and using web services.

1 Introduction

A Web Service (WS) is a programmable application logic accessible using standard internet protocols. It combines the best aspects of component based development and the web and offers functionality that can be easily implemented. Unlike current component technologies which are accessed via proprietary protocols, WSs are accessed via ubiquitous Web protocols like HTTP using universally-accepted data formats such as XML. In real business terms as Data Warehouse integrated heterogeneous data sources (base databases), WSs have emerged as a powerful mechanism for integrating disparate IT systems and assets. They work using widely accepted technologies and are governed by commonly adopted standards. WSs can be adopted incrementally with little risk and at low cost. Today, enterprises use WSs for point-to-point application integration, to reuse existing IT assets, and to securely connect to business partners or customers. Independent Software Vendors (ISVs) embed WS functionality in their software products so they are easier to deploy. From a historical perspective, WS represents the convergence between the service-oriented architecture (SOA) and the Web. SOAs have evolved over the last 10 years to support high performance, scalability, reliability, and availability. To achieve maximum performance, applications are designed as services that run on a cluster of centralized application servers. A service is an application that can be accessed through a programmable interface.

M.-C. Shan et al. (Eds.): TES 2004, LNCS 3324, pp. 136–151, 2005.

WS represents a new form of middleware based on XML and the Web and helps to solve the challenges using traditional application-to-application integration. It has several advantages over traditional middleware which, unlike WS, doesn't support heterogeneity, doesn't work across the internet, isn't pervasive, hard to use, and has high maintenance costs. WS simplifies the process of making applications talk to each other which results in lower development cost, faster time to market, and reduced total cost of ownership. Traditional middleware such as RPC, CORBA, RMI, and DCOM, relies on tightly coupled connections which is brittle and may break if any modification is made to the application. In contrast WS supports loosely coupled connections which minimize the impact of changes to applications. A WS interface provides a layer of abstraction among clients and servers and also makes it easier to reuse a service in another application whereas loose coupling reduces the cost of maintenance and increases reusability.

A WS can be developed using any programming language and can be deployed on any platform. In addition, it can be accessed by an application written in any programming language running on any platform. Although the Web supports universal connectivity, it doesn't resolve the issue of heterogeneous communication by itself. WS supports heterogeneous communication because they all use the same data format (XML), which makes it possible for communicating applications to understand each other. WSs are essentially based upon three major technologies and standards:

Simple Object Access Protocol (SOAP) provides the means for communication between WSs and client applications. It is an XML-based protocol for messaging and instead of defining a new transport protocol it works on existing protocols, such as HTTP, or SMTP. A SOAP message has a very simple structure: an XML element (the <Envelope>) with two child elements, one of which contains the optional <Header> and the other the <Body>. The <Header> contents and the <Body> elements are themselves arbitrary XML.

Web Services Description Language (WSDL) is used to describe the interfaces of a service. For WSs, SOAP offers basic communication, but it does not inform about what messages must be exchanged to successfully interact with a service. That role is filled by WSDL; an XML format to describe WSs as a collection of communication endpoints that can exchange certain messages. For developers and users, WSDL provides a formalized description of client-server interaction. Developers use WSDL documents as the input to a proxy generator tool that produces client code according to the service requirements. A complete WSDL service description provides an application-level service description or an abstract interface and the specific protocol-dependent details that users must follow to access the service at a specified concrete service endpoint.

Universal Description, Discovery and Integration (UDDI) is used to register and publish WSs and their characteristics so that they can be found by clients. The UDDI specifications offer users a unified and systematic way to find Service Providers through a centralized registry of services that is roughly equivalent to an automated online "phone directory" of WSs. On one hand, there are browser-accessible global UDDI registries available for "public" access and, on the other hand, individual companies and industry groups are starting to use "private" UDDI Registries to integrate and access to their internal services. UDDI provides two specifications which define the structure of the service registry and its operation. Registry access can be programmatically accomplished using a standard SOAP API for both publishing and querying.

However, in spite of its adaptability and scalability, it still has spatial constraints that limit its usability. We propose to eliminate most of them through significant improvements in various aspects of mobile discipline. Powerful mobile devices are becoming more common and location tracking techniques have been improved to find the location of a mobile user with precision. The widespread use of internet, electronic shopping, mobile and wireless communication systems, etc., motivated researchers and developers to migrate all types of information and also services to the web. We, therefore, envision an all powerful web system where service provisioning is also included as one of its capabilities. The migration of WSs to internet provided an excellent way to achieve application-to-application interaction which made it possible for companies to manage their business activities efficiently, economically, and with a high degree of automation. It is estimated that by 2008, the number of wireless and mobile devices will significantly outnumber the wired devices. Studies by industry analysts forecast huge demand for wireless and mobile devices, creating substantial opportunities for wireless device application and service providers. Faced with an increasingly difficult challenge in raising both revenue and numbers of subscribers, wireless carriers and their partners are developing a host of new products, services, and business models based on data services. One of the emerging technologies are the location-based services which intend to boost both service and revenue. Examples of location-based services include getting driving directions, traffic information, weather, and travel schedules, paying electronic tolls, scheduling fleets for transport operators and locating convenient modes of transportation, and locating people and businesses listed in electronic directories.

It is becoming increasingly clear that Location-based Services will play a major role in the evolution of Wireless WSs (WWS) which is the topic of this paper. In this work we develop an innovative architectural framework called *Web Bazaar* to provide access to location based WSs to all users (mobile and static) which is free from spatial and temporal constraints. Thus a user can access his desired service from anywhere and anytime in a user-friendly manner.

The rest of the paper is as follows: In section 2, we discuss previous work on WSs. Section 3 elaborates the constraints present in the existing models of WSs and their limitations in wireless and mobile environment. Discussion of the location domain and its application for development of location based WSs is presented in section 4. Section 5 explains in detail the proposed architecture of location based WWS called Web Bazaar and its working is discussed in section 6. Security of WWS and one of its possible solutions is discussed in section 7. We conclude the paper and discuss future work in section 8.

2 Previous Work

In this section we review earlier work in WSs. Mobile Resource Management (MRM) system for mobile E-commerce [9] provides location-based and Context-aware services for mobile users. It helps M-commerce service providers, such as local advertisers, to improve the effectiveness of their advertisement process and real-time E-commerce services. Ad hoc pervasive connectivity on mobile systems based on Bluetooth applications has been discussed in [1] where the proposed Ronin Agent Framework introduces a hybrid architecture that provides a simple and uniform scheme for deploying highly dynamic distributed intelligent components in a mobile world. It provides an insight into the existing discovery services like Service Location Protocol (SLP), Jini, Universal Plug and Play (UPnP) etc., and their limitations in providing Mobile E-services. It also highlights lack of rich representation, constraint specification and inexact matching and ontology support. M-commerce application of proximity based coupon delivery is discussed in [11]. In a typical scenario a merchant is notified when a valued customer is within some distance of a retail outlet, upon which the customer is delivered a coupon or some notice of a special promotion. They discussed the limitations of current architectures for providing location information, and suggested the requirements for an architecture, which would make such a service possible. The work reported in [2] proposes to define a caching service for WSs in mobile wireless ad hoc networks. The work is based on the previous experiments where authors have explored caching CORBA - based services as a way to increase their availability, predictability, times responsiveness and scalability. The author identified the maintenance of original programming model and sufficient transparency to be a big challenge because the interceptors are not yet standardized in WSs. Two scenarios of using WSs in mobile devices have been described in [8]. The first scenario considers mobile devices acting as service requestors. It is most suitable for nomadic users who try to locate a product or service close to their physical location to manage their personal information hosted on a central server and to administer their events by invoking relevant WS. In the second case a proxy plays the role of the mobile representative in the fixed network architecture. This proxy interacts via WS-aware protocols with the service broker and the service provider and returns the results to the mobile devices using WS-agnostic protocols such as WAP/WML over a wireless network. In the AROUND architecture [3], two distinct models (distance and scope based) of location based [7, 9, 10] selection have been considered.

In the distance based model clients select the servers located within some distance from their own position. The main limitation of this model is that the correlation between context and proximity tends to decrease as the notion of proximity is enlarged. In the scope based model each service has a service scope that explicitly represents the usage context of that service as a region in physical space. The clients select only those services that match their location. The distance based model places the focus on the location of the server providing the service whereas the scope based model places the focus on the geographical area defined for the service usage.

The resource demands for environment that wish to support mobile code to enable disconnected operation have been examined in [4] and some existing commercial service discovery mechanisms have been discussed in [5, 6]. Authors argued that DAML (DARPA Agent Markup Language) was expected to change the way people and machines browse the Web. They also try to show that it can be used to change the way services are described and discovered in the wired and the wireless world.

The WSs caching for seamless access in the event of disconnections is investigated in [13]. The author suggested that continued access to WSs from mobile devices during disconnections could be provided by a client-side request-response cache to a limited extent. The Cool town project at HP Laboratories [12] addresses the problem of service discovery for nomadic computing.

The MUSA-Shadow project [14] aims to avoid the fragmentation of the web into spaces that are solely accessible with specific type of devices while providing an extensible and flexible infrastructure for location based services. The issues concerning the development of an extensible platform for location based services (LBS) provisioning was presented in [15]. Authors focused on the exploitation of XML either through PoLos specific syntaxes or through WS interfaces. In [16] authors proposed to advance the art of location based applications thus enhancing end-user applications and new commercial opportunities.

3 Limitations of Web Services in Wireless and Mobile Environment

The present model of WSs has some fundamental limitations which affect their seamless infusion into the wireless and mobile environment. This section reviews some of them and discusses the need for location information in the existing structure to make it commercially useful.

- **Service discovery:** In current WS infrastructure there is no support for location dependent service discovery. A user, when searching for a service, has to scan the entire range of available services published in the UDDI. This not only is inefficient but consumes more power which is not acceptable to mobile systems. A location based WS system can immensely enhance the efficiency of a system. The user will get a more concise or filtered view of the available WSs according to his choice of location. For example, if a user wants to book a hotel in the Plaza area of Kansas City, then his request should be responded with only those hotels which are located in and around Plaza.
- **Pull based information retrieval:** WSs support request/response style of messaging. In wireless environment with large mobile user population this style of information access may suffer from scalability bottleneck as the number of users' request increases. An amicable solution would be to broadcast the information of the WSs. Thus, publishing WSs by broadcasting them to the user can reduce the request portion of information exchange in a mobile WS system. We know that any action that makes a WSDL document available to a service requestor, at any stage of the service requestor's lifecycle, qualifies as service publication [17]. We argue

that broadcasting WSDL [18] documents containing description of WSs is a convenient way of publishing WSs and can be categorized as direct publication.

- **Personalization of WSs:** To further improve the availability of WSs, all location dependent services should be complemented with personalization. Personalization in location-dependent services is a good way of improving the usability of the services by providing the essential and probable information. It is a big challenge to design personalized location-awareness and location dependency so that it does not require too much effort.

- **Topical WSs:** WSs which handles topical data may be categorized as Topical WSs. Topical information is important to users. This is the kind of information that may change while the user is on the move, in which case the information previously checked or accessed may no longer be valid. Examples of such topical infor- mation are traffic information, weather forecasts, last minute theatre ticket deals, stock information and foreign exchange rates, flight bookings, etc. An important issue is whether the user needs the information when he or she is at the given location, before getting at the location, or when planning a visit. Flexibility in mobile environment requires that one should support both pre-trip planning and on-route information about occasionally found points of interest. The user should face least constraints when availing these services. At present, no existing architecture handles the issue of Topical WSs separately.

- **Limitations of UDDI:** At present UDDI [17] does not support semantic description of services and since it depends on the functionality of the content language, it is very difficult to automate service discovery to get accurate information. It provides keyword-based search. For example if restaurant in Chicago is queried, the result would also include restaurants with names like Chicago Uno which is in Kansas City. It is logically centralized but physically distributed repository of WSs. Currently UDDI consists of three components (a) white pages of company contact information, (b) yellow pages that categorize businesses by standard taxonomies and (c) green pages which document the technical information about services that are available. This structure of UDDI will not be viable for discovering services especially if we consider location based services like local traffic information, local weather, and so on.

4   Location Based Web Services

Location based WSs are an important class of context aware applications. We argue that incorporating location information in a WS can significantly decrease the service discovery time on part of the user. This property becomes more significant for a user with a mobile unit which has limited storage and processing capability. The mobile unit when searching for a service in a particular location expects to discover services which are in and around the desired location. There are a variety of applications like traffic information, restaurant and hotel booking, serendipitous location based search of public places like fast food joints, gas stations, post office, grocery stores, etc., which, if made location dependent, can make a strong impact in our day to day lives.

There has been some work to represent locations. We present a brief description of two of the most prominent existing location models [19].

- **Geometric Model:** In this model, the location is represented by sets of coordinated n-tuples. This model is based on reference coordinate systems, the coordinates of which are returned by the Global Positioning System. One of the advantages of this model is the accuracy with which it locates an object. Another advantage is that it is compatible across heterogeneous systems. But it can be very costly and complex to implement and thus may not be suitable for mobile units. Further, GPS signals are weak inside buildings which make it ineligible for use in every environment.
- **Symbolic Model:** This model resembles the real world entities like cities, streets and buildings. This model is simple to handle and has coarser granularity than the geometric model as it is based on the relations between the real world objects. Further, less amount of data is involved which makes it easier to manage. Thus, this symbolic location model seems more suitable for mobile computing environments. One of the disadvantages is that it is difficult to convert location information across heterogeneous environments.

The Universal Location Domain (ULD): Our approach is to create a framework which is compatible across all platforms. To achieve this, we propose to create a Universal Location domain (ULD). The ULD contains locations which are hierarchically arranged in a structure called the location tree. This idea is motivated by several facts. To provide ubiquitous computing ability, WSs should have the compatibility across all types of mobile devices and across all types of platforms. Moreover, Service Providers are not unique across different parts of the world and only the presence of a unique location structure may proliferate the use of location dependent WSs. We define a location in the ULD as follows:

Location: A location is a symbolic representation of a real and physical space which is designated by Cartesian Coordinates. A location is non-atomic in nature and it has the possibility of being distributed into newly constituted locations.

The location tree is a set of locations arranged in a hierarchical manner. An important property of the locations present in the ULD is *Containment*. The containment property helps to determine the relative position of an object by defining or identifying locations that contains those objects. The subordinate locations are hierarchically related to each other. Thus, Containment property limits the range of availability or operation of a service. We use this important property in location dependent WS discovery.

Apart from the ULD, we also define the location of the WS which is provided by the user who intends to access the service. It is the job of the location framework to create a location structure for the service. This location structure is then suitably mapped on the ULD to find the exact location of the WS. Consequently, the requested service is provided to the user. We present an example to explain the functionality of the proposed location framework.

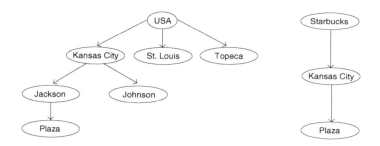

Fig. 1. A Portion of (ULD)   **Fig. 2.** Location Structure of WS

A mobile user intends to access the WS of the Star Bucks Coffee Shop and place an order for a home delivery. He initiates the order by placing a query for the availability of his choice of coffee in the Star Bucks Plaza branch of Kansas City. It is the job of the location framework to develop the location structure of the requested WS. Here, the location structure of the WS as depicted in Fig. 1 is: Star Bucks → Kansas City → Plaza. To search and access the required WS, a unique ID for the WS has to be generated. This unique ID is generated by mapping the location structure of the WS with the ULD. The location of the WS in the ULD is: USA → Kansas City → Jackson → Plaza (Fig.2). The Containment property limits the search of the location of the various Star Bucks services which are present only at the subordinate locations of "Plaza". The number of searched entries for the desired WS is equal to n where n is the number of subordinate locations near Plaza. Thus, the set of location IDs of the requested WS is: Star Bucks → USA → Kansas City → Jackson → Plaza → X_i where i is the number of subordinate locations in "Plaza". It is possible that the Star Bucks coffee shop is not present in all the subordinate locations of the "Plaza". Thus, the number of results will always be greater than or equal to the number of original locations of the WS. The exact number of locations will be found when calculated IDs will be matched with the entries of the WS present in the Distributed Service Repository (DSR), a structure which we discuss in our proposed architecture of Web Bazaar in the next section.

5   Web Bazaar: A Broadcast Based Location Dependent Web Service Architecture

Our approach attempts to address some of the limitations of WSs in wireless and mobile environments discussed in Section 3. The Web Bazaar architecture consists of a Broadcast Server, uplink and downlink Wireless Channels, Distributed Service Repository (DSR) containing Service Marts and the Universal Location Domain. The broadcast server has the central role of service distribution. In this model we propose to broadcast the services instead of the traditional pull based access. We argue that in the future, to make the WSs popular among the increasing group of mobile users, efforts should be made not only to publish the services but also to distribute them to the users. This will support both pre-trip planning and on-route information on occasionally found points of interest. Moreover, as discussed in Section 3, WS use XML documents the size of which tends to be much larger than traditional text messages. Thus, efforts should be to the minimize number of messages exchanges between a mobile user and the WS provider. Broadcasting WS information may significantly reduce the number of messages required for the process of service discovery. This is our motivation behind the proposition of WS broadcast in Web Bazaar.

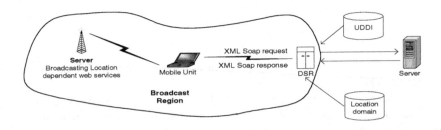

Fig. 3. Web Bazaar: Wireless Broadcast based Location dependent WS Architecture

Fig. 3 shows the proposed architecture of Web Bazaar. The standards and specifications demand that the structure and role of UDDI should not be changed. Thus, to incorporate location information, we propose the concept of Distributed Service Repositories (DSR). A DSR contains WSs entries which are local to the region. Each broadcast region contains a DSR. The DSR incorporates the location information in the WSs.

For a particular broadcast region, the WSs which cater to the region are extracted from the UDDI. The location ID for the WS is generated by using the Universal Location Domain. For any WS located at a particular location, its location structure is generated and mapped with the ULD to get the unique location ID. This means for a single WS, there will be many entries depending on the number of locations at which the service is located. This may increase the size of the DSR, but considering the fact that the DSR contains entries which are local to the broadcast region, the number of entries will be limited. A different view of the WS is then generated. This view contains fewer parameters which makes it suitable for broadcast in bandwidth limited channels. Views are compact documents and contain information about the location dependent WSs. The views are broadcasted and used to initiate the service from a mobile or wireless device.

Within a DSR, the services may be organized by keeping similar WSs under one group. For example, the DSR in Kansas City will group the WSs related to hotels under one group, grocery stores under another group, and so on. Each specific group is called a Service Mart. This makes the service discovery much easier. Moreover, It helps in creating simple but informative views. For example, if a user in Lenexa intends to use the WS of grocery store in downtown area in Kansas City, he just have to give the location name and type of Service Mart (here, for example, Food Mart) in his request to discover a service. In contrast to the earlier method of searching the UDDI registries and then deciding for a service, our way of service discovery through broadcast of views seems much faster. As evident from table, these views are much simpler than the UDDI entries.

Table 1. Views of two Types of WSs

Service Name Mart Type	Input, Output Parameters	Location
Star Bucks Fast Food	CoffeeType, Order Destination, Card No.	Plaza
Theatre Ticket Entertain	Movie, Showtime,No.ofTickets,Card No.	Down Town

The advantages of creating DSRs are manifold. First, it helps in creating location dependent WSs by assigning unique location IDs to each service. This helps in fast service discovery. Second, the views created for the WSs are compact so that they may be used to broadcast the service definitions in bandwidth limited wireless channels, thus supporting our broadcast mechanism. Third, a user does not have to contact the UDDI for service discovery for location dependent services. By mentioning the location, the request can bypass the UDDI and contact the corresponding DSR directly. This prevents exhaustive search, allows fewer data download which is suitable for bandwidth limited wireless environment and allows fast access to the service.

There is an important issue which needs to be discussed at this point. There is a large number of WS entries present in the UDDI. Broadcasting all of them through bandwidth limited wireless channels may sound unrealistic at present. At the same time we argue that broadcasting location dependent WSs present in DSRs which are commonly used in day to day life will certainly make them more popular among the mobile users. Prominent among these services are the Topical WSs in which the information accessed frequently change when the user is on the move. Important examples are flight bookings, last minute theatre ticket deals, traffic information etc.

6   Working of Web Bazaar

The broadcast mechanism consists of a broadcast channel from the broadcast server to the users. It also consists of an uplink channel from the users to the DSR and a downlink channel from the DSR to the user (Fig.3). Each broadcast is preceded by an index which depicts the sequential order of WS broadcast. The index is also interleaved between the broadcast views so that the user does not have to wait for the next broadcast schedule. The structure of the index will be helpful to the users in personalization of WSs which we discuss later. The broadcast includes the compact views which contain WSDL definitions. These definitions are WSDL components. The WSDL components consist of interfaces, bindings, and services.

The service download, request and response of services are managed by a Java based Coordinator present in the mobile device. This Java based Coordinator is installed in the machine when the user avails the Web Service from the service provider. The Coordinator has the task of listening to the channel, downloading the required service, initiating the request, and receiving the response. The Coordinator also has an important task of Personalization of the WSs.

Personalization of WSs means to access and use Services only according to a pre-planned schedule fixed by the user. The Coordinator downloads the index containing the description of the services to be broadcast. The user checks the index according to his needs and so does not have to bother about the downloading of the services. Based on the Service checked and the predetermined schedule from the index, the Coordinator estimates the time required for service download. It allows the mobile device to go into doze mode to save power. It becomes active only when the service is about to be broadcasted. The service components are downloaded and the user is notified to start the service request. Thus, only those services needed by the user are downloaded from the channel. This describes the ability of the user to block certain services data and personalize his requirements.

To initiate a service request, the Coordinator creates a SOAP message. The XML document in the message is created according to the downloaded service description which contains the unique location ID (for a location dependent service) and user's ID. Even if the view of the WS of the desired location is not present, their structures and properties of XML documents allow changing the view information to access a particular location dependent service.

Example:

Star Bucks	Fast Food	**CoffeeType 1**,Order Destination, Card No.	**Plaza**

⬇

Star Bucks	Fast Food	**Coffee Type 2**,OrderDestination, Card No.	**Walnut**

If the user wants to order coffee of type 2 which is not available in the Plaza branch, he doesn't have to wait for the view of Walnut to be broadcasted. He can simply change the XML information of the views by replacing the *Coffee Type* Information and the *Location* information. It is the job of the ULD to map the location information provided by the user to generate appropriate location IDs which can be used to search the appropriate location dependent service from the DSR.

After XML request is constructed, the coordinator sends this message to the SOAP server present in the DSR through the uplink wireless channel. Since this message is compact, it is the responsibility of the DSR to make it compatible for the Web Server. When all the definitions are added, The SOAP server directs the request to the service provider's WS. If the user requests a service which belongs to another location, the request is transferred to the DSR containing the service description. This identification of location is done by the DSR local to the user.

The WS, after receiving the requests, processes it and creates a response which is also a SOAP message. It is sent to the DSR local to the user. The DSR operates on the response and makes it compact thus making it easier for the DSR to push the response through the downlink channel to the user. The compaction is necessary as the SOAP messages which are several times larger than text messages may overburden the bandwidth limited wireless channels. When the response reaches the user, he is notified about its arrival.

The proposed framework thus provides location dependent WS to the mobile user. Broadcasting WS information signifiicantly reduces the number of messages in the wireless environment. The broadcast XML views are compact and allows efficient service request/response style of messaging. The ULD is used to create appropriate location information for the service entries in DSR and also for service request. The simplified hierarchical structure of the ULD allows smooth addition/deletion of location information when needed.

7 Security of Wireless Applications

Wireless systems face greater security risk than wired systems and this is true also for WWS. WWS exposes companies to a massive range of new threats and vulnerabilities. The overall security of a wireless application is only as strong as its weakest link, and in a mobile-commerce network, the weakest link is the mobile device. We discuss the risks and security threats in a service framework. It can be inferred that existing wireless security controls are inadequate to deliver the levels of security that the next wave of WWS will demand. Security issues related to communication, theft, attack, DoS (denial of services), etc., are important but we only propose to investigate security issues related to user authentication. Some of the key security issues with wireless application security include:

- **Confidentiality:** Access to confidential and sensitive data should be restricted to only those authenticated users.
- **Availability:** Mission-critical data and WSs should be available with contingency plans to handle catastrophic events such as infrastructure failures, security breaches, etc.
- **Integrity:** The integrity of data transmitted over wireless network from the point of transmission to the point of delivery needs to be extremely well maintained.
- **Privacy:** Wireless providers should take care to adhere to the legal requirements to safeguard user privacy. This is particularly significant for location-based services as there is an inherent possibility that the users can be tracked. However, the availability of location information can be turned into a security advantage.

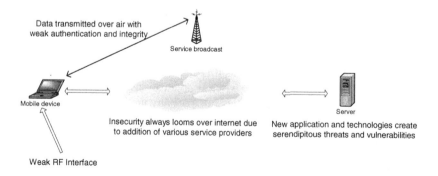

Fig. 4. Insecurity in Wireless Services

Figure 4 depicts the security threats faced by the Wireless Services. We describe the security needs of WWS with the following list and then present an idea of our scheme to protect authorized users from serendipitous threats and malicious attacks.

- Since the mobile device has a weak radio frequency interface, data transmitted over wireless networks such as passwords, personal information, security information, etc. can be captured using digital RF scanning equipment. Most wireless protocols do not come with built-in encryption mechanisms. Additional security measures such as secure connections and cryptography are definitely needed, especially for those applications transmitting sensitive data.
- Data transmitted over air has weak authentication and integrity thus allowing evesdroppers to manipulate data and control information.
- Insecurity always exists over the internet due to incessant additon of large user base and various service providers.
- As new Application to Application (A2A) integration increases, they create serendipitous threats and new vulnerabilities to the existing WS Systems.

Location Signature Based Security and Authentication:

One of the biggest obstacles to the widespread adoption of WWS is winning the trust of mobile users. A single security breach provides a very high profile way of undermining a wireless service. We propose the use of "Symbolic Location Coordinates" identifying the real time location information of a user into existing security mechanisms to improve the efficacy of authentication, authorization, and access controls. We refer to this real time location information which will be unique for a user as "Location Signature (LS)".

Effective wireless application security depends on the ability to authenticate users and grant access accordingly. Existing authentication and authorization mechanisms fundamentally depend on information known to a user (password or keys), possession of an authentication device (security token or crypto card) or information derived from unique personal characteristics (biometrics). None of this is totally foolproof. Symbolic Location information (building, street, area ID, base station id, etc.) of a mobile device or user adds a fourth and new dimension to wireless application security. It gives extra assurance to users of the wireless applications who want to perform sensitive operations such as financial transactions, access valuable information, or remotely control critical systems. It can supplement or complement existing security mechanisms. The Location Signature can still be used as a security mechanism when other systems have been compromised as it is and will always be unique for a user at any point of time. For highly sensitive wireless applications, a real time Location Signature can be generated so that authorities can trace any malicious activity back to the location of the intruder. Without the incorporation of LS, it will be difficult to trace the origin of any malicious activity.

It is almost impossible to replicate a Location Signature because a user cannot exist at two places at the same time and use it elsewhere to gain unauthorized entry. Even if the information is intercepted during communications, an intruder cannot replicate that data from some other place. A Location Signature is continuously generated from location information on real-time basis and is unique to a particular place and time. Such information becomes invalid after a short time interval, which means that the intercepted Location Signature cannot be used to mask unauthorized access especially when it is bound to the wireless protocol messages as checksums or digital signatures. Even if the perpetrator uses other means to masquerade as a legitimate user, the complete set of Location Signatures can be used to log the access trail [20].

We propose to develop and incorporate two-way authentication between the wireless client (mobile device) and the Web Server. The Web Server can give access to a wireless client based on the security mechanisms along with the LS. The reverse process - the client receiving Server's LS - ensures a higher lever of security as it will always be a unique mapping between the continuously moving user and the Web Server, especially if this "handshaking" is done periodically. This requires an additional message between the wireless device and the Web Server which is not likely to affect the performance. Location information can provide evidence to absolve innocent users. If illegal activity is conducted from a particular user account by someone who has gained unauthorized access to that account, then the legitimate owners of the account might be able to prove that they could not have been present in the location where the activity originated.

The following example gives an idea of the working of WWS based on LS:

Suppose Ben is on his way back home to Overland Park and wants to buy coffee from Star Bucks. He is tired and in no mood to search the nearest Star Bucks on his way back. He uses his PDA and logs on to the Wireless Broadcast of WWS. He downloads the views of Star Bucks and places the order. He types the necessary information about the type of coffee he wants, the address of order delivery and initiates a query. The mobile device coordinator of Ben's PDA develops the location information for delivery based on the address. It also develops the location signature and attaches with the query. The LS consists of the mapping among various real time variables like the time, building, street, area ID, mobile device ID, etc. After the LS is developed, it is attached with the query and also stored in a log file corresponding to the time when the signature was generated. The log file is not accessible to the user. This LS is a real time variable (as the values of most of the variables changes every moment) but will always be unique for a particular time. It is also cumulative, i.e., the new signature is a set of old signatures plus the new signature recently added. The Web Server receives the views through the wireless channel and based on the location finds out the service for Star Bucks which is nearest to Ben's preferred location of Overland Park. It sends back the response and asks Ben to place the order. Ben enters the necessary information like his credit card number and the address for delivery and completes the order. When the server receives the information, it checks the log file and matches the existing LS set with the newly received LS set. If the user is original and not a malicious one, both the LS set will exactly match except the last entry which is newly generated. The two sets will never match for a malicious user. The Web Server completes the transaction and issues a response. This response will be available to the user only when the received location signature set matches exactly with the signature set present in the mobile device cache.

Thus, location-based authentication can be done transparently to the user and be performed continuously. This means that unlike most other types of authentication information, LS can be used as a common authenticator for all systems the user accesses. This feature makes location-based authentication a good technique to use in conjunction with single sign-on.

8 Conclusion and Future Work

In this paper we have proposed a wireless broadcast based WS architecture called Web Bazaar in which a Universal Location Domain is used to develop unique location ID for a service. The unique location ID is subsequently used for fast and efficient service discovery. The wireless broadcast consists of the views of WSs. These views are compact XML documents and they can be efficiently used to decrease the message exchange between the mobile user and the web server which is essential in a bandwidth limited wireless environment. Security of wireless applications is a huge concern as the mobile unit has weak radio frequency interface. Moreover, increase of application to application (A2A) integration in WSs introduces unexpected security lapses. To prevent this, we have proposed a location based signature scheme which will add another dimension to the existing set of authentication mechanisms. In our future work, we intend to develop the location structure further by introducing a location domain naming system similar to the existing Domain Name Services (DNS) for translating domain names of IP addresses. This will enable simpler and more realistic location identification for location dependent services. The location domain naming system may also be used for the location signature scheme. We also intend to develop the algorithm for the creation of location signature and authentication procedures of the location based signature scheme.

References

1. Hary Chen, Anupam Joshi, Tim Fini, Dynamic service Discovery for mobile computing: Intelligent Agents meet Jini in the Aether, cluster computing volume 4,issue 4, Special issue on internet scalability.
2. Roy Friedman Caching Web services in Mobile Ad-hoc Networks: Opportunities and Challenges, POMC'02 October 30,31,2002, Toulouse, France.
3. Rui Jose , Adriano Moreira and Helena Rodrigues, The Around architecture for Dynamic Location based services, Mobile Networks and Applications, 8, 377-387,2003.
4. Bharat Chandra, Mike Dahlin, Lei Gao, Amjad Ali Khoja, Amol Nayate, Asim Razzaq, Anil Sewani, Resource Management for Scalable disconnected Web services, WWW10, May 1-5, Hong Kong.
5. Dipanjan Chakraborty, Filip Perich, Sasikanth Avancha, Anupam Joshi, DReggie: Semantic Service Discovery for M-Commerce Applications, Workshop on Reliable and Secure Applications in Mobile Environment, 20th Symp. Reliable Systems, 2001.
6. Anupriya Ankolenkar, Mark Burstein, Jerry R. Hobbs, Ora Lassila, David L. Martin, Drew McDermott, Sheila A. McIlraith, Srini Narayanan, Massimo Paolucci, Terry R. Payne and Katia Sycara, "DAML-S: web services description of Semantic web". Proc. 1st Int. Semantic Web Conference.
7. Upkar Varshney, "Location Management Support for Mobile Commerce Applications". In Proceeding of workshop WMC '01.

8. T. Pilioura, A. Tsalgatidou, S. Hadjiefthymiades, "Scenarios of using Web Services in M-Commerce", ACM SIGecom Exchanges, Vol. 3, No. 4, January 2003.

9. Olga Ratsimor, Vladimir Korolev Anupam Joshi, Timothy Finin, "Agents2Go: An Infrastructure for Location-Dependent Service Discovery in The Mobile Electronic Commerce Environment". Proceedings of the WMC'01.

10. Lai Jin, Tatsuo Miyazawa, "MRM Server: A Context-aware and Location-based Mobile E-Commerce Server". In proceeding of workshop WMC '02 pages 33-39.

11. Jonathan P. Munson, Vineet K. Gupta, "Location-Based Notification as a General-Purpose Service". In proceeding of workshop WMC 02.

12. Tim Kindberg and John Barton: A Web-based nomadic computing system, Computer Networks 35(4), March 2001, pp 443-456.

13. Stans Kleijnan and Srikanth Raju, An Open web services Architecture, Sun Microsystems, INC.

14. Sebastian Fischmeister, Guido Menkhaus Wolfgang Pree MUSA-Shadow: Concepts, Implementation, and Sample Applications: A Location-Based Service Supporting Multiple Devices: 40th International Conference on Technology of Object-Oriented languages and systems, Sydney, Australia, 2002.

15. Anastasios Ioannidis, Manos Spanoudakis Panos Sianas Ioannis Priggouris Stathes Hadjiefthymiades Lazaros Merakos, Using XML and related standards to support Location Based Services, SAC 04', March 14-17,2004, Nicosia, Cyprus.

16. Chatschik Bisdikian, Jim Christensen, John Davis II, Maria R. Ebling, Guemey Hunt, William Jerome, Hui Lei, Stephane Maes, Daby Sow, Enabling Location-Based Applications, WMC 01 Rome Italy.

17. UDDI, www.uddi.org

18. WSDL, www.w3c.org/TR/wsdl

19. Dik Lun Lee, Jianliang Xu, Baihua Zheng, Wang-Chien Lee, Data Management in Location-Dependent Information Services, IEEE Pervasive Computing, July-September 2002(Vol 1, No.3).

20. Harsha Srivatsa, IBM developer works, Location-based Security for Wireless Applications, November, 2002.

Financial Information Mediation:
A Case Study of Standards Integration for
Electronic Bill Presentment and Payment Using
the COIN Mediation Technology

Sajindra Jayasena[1], Stéphane Bressan[2], and Stuart Madnick[3]

[1] Singapore-MIT Alliance
Sajindra@mit.edu
[2] School of Computing, National University of Singapore
steph@nus.edu.sg
[3] Sloan School of Management, Massachusetts Institute of Technology
smadnick@mit.edu

Abstract. By its very nature, financial information, like the money that it represents, changes hands. Each player in the financial industry, each bank, stock exchange, government agency, or insurance company operates its own financial information system or systems. Therefore the interoperation of financial information systems is the cornerstone of the financial services they support. E-services frameworks, such as web services, are an unprecedented opportunity for the flexible interoperation of financial systems. Naturally the critical economic role and the complexity of financial information led to the development of various standards. Yet standards alone are not the panacea: different groups of players use different standards or different interpretations of the same standard. We believe that the solution lies in the convergence of flexible E-services such as web-services and semantically rich meta-data as promised by the semantic Web; then a mediation architecture can be used for the documentation, identification, and resolution of semantic conflicts arising from the interoperation of heterogeneous financial services. In this paper we illustrate the nature of the problem in the *Electronic Bill Presentment and Payment* (EBPP) industry and the viability of the solution we propose. We describe and analyze the integration of services using four different formats: the IFX, OFX and SWIFT standards, and an example proprietary format. To accomplish this integration we use the COntext INterchange (COIN) framework. The COIN architecture leverages a model of sources and receivers' contexts in reference to a rich domain model or ontology for the description and resolution of semantic heterogeneity.

1 Introduction

Effective and transparent interoperability is vital for the profitability and sustainability of the financial Industry. Adhering to a standard is not feasible because different institutions often utilize different standards. Even when within one standard

M.-C. Shan et al. (Eds.): TES 2004, LNCS 3324, pp. 152–169, 2005.

or when standards seem to agree, one often finds different possible interpretations originating in the particular practices and cultural background of the various players.

Typically, a Financial Institution (FI) that is involved in *Electronic Bill Presentment and Payment* (EBPP) Industry, for instance operating in a European Union country, is faced with a multitude of standards such as IFX (Interactive Financial Exchange protocol)[10], OFX (Open Financial Exchange Protocol)[9] and the world wide inter-bank messaging protocol, SWIFT [11]. Making matters worse, the FI may have its own semantics for its internal information systems that represent the same business domain in a different context. In the rest of this paper we would be referring to the set of assumptions about the representation, syntax and interpretation of data according to IFX, OFX, and SWIFT as IFX, OFX and SWIFT contexts and the assumptions of the internal financial system of a Financial Institution as internal context.

The *Price* and *Invoice* concepts may be represented in different ways, e.g., excluding tax, with tax and fees, and even with inter-bank charges, resulting in definitional conflict [1]. Interoperability of such definitional conflicts is vital in distinguishing intra-bank and inter-bank payment across borders. Further, different contextual heterogeneities exist on the *currency*, where in certain contexts like IFX and OFX it is implicitly based on where the funds are directed. As a result of different *Account types* and BANK/*BRANCH code*, financial institution would need to maintain complex mappings between different contexts. In addition, there can be data level heterogeneities like date formats and representations. Examples of possible conflicts are summarized in Table 1. The columns for OFX, IFX, and SWIFT represent actual real-life conflicts and similarities that exist between those standards, while the conflicts addressed under the internal schema column refer to a hypothetical, but realistic, financial system that would interact between OFX, IFX and SWIFT standards.

Table 1. Some Conflicts in Different Standards

Property	Internal Schema	OFX	IFX	SWIFT 103/103+
Price	1000 FFR (French Franc)	1000 USD + 1000 * 5%	1000 USD + 1000 * 5%	1000 USD + 1000* 5% + 10 USD (inter-bank charge if outside EU)
Currency	FFR	Currency of country of incorporation of payee bank i.e. USD	Currency of country of incorporation of payee bank i.e. USD	Specified in message – can be the payee or payer's currency
Account types	CHK,SVG, MNYMRT	CHECKING, SAVINGS	DDA,SDA	N/A
Bank and branch code	Internal ID	Dependent on the country i.e. clearing #,sort #	Dependent on the country i.e. clearing #,sort #	BIC / BEI (branch ID + bank Id)
Invoice	Net	Net + fees + tax	Net + fees + tax	Included in *Amount* – N/A
Due date	*23022002*	*20020223*	*2002-03-23*	*030223*

The objective of this research is to analyze how COIN mediation technology [2, 3, and 8] could be applied to provide a declarative, transparent yet effective mediation solution to the sources of heterogeneity and conflicts that exist within and among the existing financial standards. Further we discuss the extension of our work in mediating the conflict that cannot be addressed in the current COIN implementation.

The organization of the following sections is as follows. First we look at the plethora of financial messaging standards that infest the financial world followed by a review of related work in mediation technologies and specific work related to interoperability in the financial industry. Then we look at the intricate details of the COIN mediation framework. Next, the bulk of the paper focuses on how COIN can be applied in one of the critical industries in the financial world – The *Electronic Bill Presentment and Payment* (EBBP) industry. In the final section, we summarize and briefly discuss future research.

2   Background and Related Work

2.1   Financial Standards

The standards addressed herein are involved in business banking, Electronic Bill Presentment and Payment, Securities and Derivatives, Investments, Economic and Financial indicators, straight through processing and other *over the counter* derivatives. As a whole, the financial industry is cluttered with numerous protocols and standards that are utilized in different segments in the financial industry. Prominent ones are Financial Information Exchange protocol (FIX), S.W.I.F.T., Interactive Financial Exchange (IFX) and Open Financial Exchange (OFX). SWIFT is the leader in inter bank transactions, and also has gained a significant market holding on Securities and derivatives, payments as well as investments and treasury after introducing a set of messages for securities and derivatives industry. OFX is the leader in Intra-bank transaction systems followed by its successor, IFX. IFX is opting to replace OFX, through its rich and extended messaging standards. Both of these standards are widely used in business banking, Electronic Bill Presentment and Payment, ATM/POS Industry. FIX is the leader in securities and derivatives market, used by major stock markets around the world. Most of these protocols use XML as the medium of messaging. Non-XML based standards like FIX and S.W.I.F.T have come up with XML versions, namely *FIXML* and *'SWIFTStandards XML'*. In addition to these major players, some of the other protocols are *RIXML* – Research Information exchange and *IRML* – Investment research markup , focusing on fixed income securities and Derivatives market, *MDDL* - Market Data Definition and *REUTERS* in economic and industrial indicators, *STPML* – Straight through processing markup language - a superset protocol to replace *FIX,SWIFT ISITC* and *DTC ID, FinXML* – Financial XML which focuses on Capital market instruments and straight through processing (STP) and finally *FpML* - Financial products markup language focusing on interest rate swaps, forward rate agreements, Foreign Exchange and other *over the counter* derivatives.

2.2   Different Mediation Strategies

When institutions exchanging financial information subscribe to different standards, a mediator can be used to translate from one encoding scheme to another. The problems that the mediator needs to solve are similar to those in data integration of heterogeneous sources, where the potential variety of encoding schemes can be arbitrarily large in the latter case.   The approaches addressing the issue of interoperability of disparate information sources have been categorized in the literature as static vs. dynamic [14], global vs. local schema [15], and tightly vs. loosely coupled [16, 17] approaches. These groupings can roughly be thought of referring to the same distinction characterized in [16] by:

- Who is responsible for identifying what conflicts exist and how they can be circumvented; and
- When the conflicts are resolved.

We briefly look at these approaches under the categories of tightly and loosely coupled approaches.

In tightly coupled approaches, the objective is to insulate the users from data heterogeneity by providing a unified view of the data sources, and letting them formulate their queries using that global view. In *bottom up* approaches the global schema is constructed out of heterogeneous local schemas by going through the tedious process of schema integration [18]. In *top-down* approaches global schema is constructed primarily by considering the requirements of a domain, before corresponding sources are sought. In tightly coupled approaches, data heterogeneities between sources are resolved by mapping conflicting data items to a common view. Early prototypes which have been constructed using the tight-coupling approach include Multibase [19], ADDS [20], and Mermaid [21]. More recently, the same strategy has been employed for systems adopting object-oriented data models (e.g. Pegasus [22] based on the IRIS data model), frame-based knowledge representation languages (e.g. SIMS [17] using LOOM), as well as logic-based languages (e.g. Carnot [23] using CycL, an extension of first-order predicate calculus).

Loosely coupled approaches object to the feasibility of creating unified views on the grounds that building and maintaining a huge global schema would be too costly and too complex. Instead they aim to provide users with tools and extended query languages to resolve conflicts themselves. Hence, instead of resolving all conflicts *a priori*, conflict detection and resolution are undertaken by receivers themselves, who need only interact with a limited subset of the sources at any one time. MRDSM [15] is probably the best-known example of a loosely-coupled system, in which queries are formulated using the multidatabase language MDSL. Kuhn et al [24] have implemented similar functionalities in VIP-MDBS, for which queries and data transformations are written in Prolog. They showed that the adoption of a declarative specification does in fact increase the expressiveness of the language. Litwin et al [25] has defined another query language called O*SQL which is largely an object-oriented extension to MDSL.

In the past two decades, various mediation strategies have been developed attempting to tackle these semantic heterogeneity problems. We will not give a detailed review of these approaches; interested readers are referred to [29, 30, 31] for recent surveys. For example, the authors of [28] use a domain model and source modeling to realize and optimize queries to distributed and heterogeneous sources. Generally, under these strategies, the mediator needs to be rebuilt when the underlying sources or user requirements change, which hinders the extensibility of the approach. We will discuss a middle ground approach that overcomes these drawbacks in Section 3.

2.3 Current Integration and Mediation Strategies in Financial Standards

Due to intricacies and inefficiencies in using and integrating multiple standards and additional overhead, financial institutions as well as government organizations have put effort in merging different standards or coming up with new super set standards to replace the existing diverse standards.

One example is the effort by FIX, SWIFT, OPEN APPLICATIONS GROUP and THE TREASURY WORKSTATION INTEGRATION STANDARDS TEAM (TWIST) to outline a framework of cooperation and coordination in the area of the content and use of a core payment kernel XML transaction.

Also the Organization for the Advancement of Structured Information Standards (OASIS) is carrying out research on one XML based super set protocol that would cover all business areas. But all these effort are focused on futuristic direction rather than the problem at hand. The effort of migrating the diverse world-wide standard to a common standard would be an enormous task. Current business integration efforts like the Microsoft[TM] BizTalk Server support diverse messaging standards integration through its rich messaging and mapping framework, but lack the sophistication in mediating complex ontological and contextual heterogeneities.

3 The COntext INterchange (COIN) Approach

The COntext INterchange (COIN) framework is neither a tightly coupled nor a loosely coupled system; rather, it is a hybrid system. It uses a mediator-based approach for achieving semantic interoperability among heterogeneous information sources. The approach has been detailed in [8]. The overall COIN approach includes not only the mediation infrastructure and services, but also wrapping technology and middleware services for accessing source information and facilitating the integration of the mediated results into end-users' applications. The set of context mediation services comprises a context mediator, a query optimizer, and a query executioner.

The context mediator is in charge of the identification and resolution of potential semantic conflicts that exist in a query. This automatic detection and reconciliation of conflicts present in different information sources is made possible by general knowledge of the underlying application domain, as well as the informational content and implicit assumptions associated with the receivers and sources.

The declarative knowledge is represented in the form of a domain model, source descriptions, a set of elevation axioms, a set of context definitions, and a conversion library. The result of the mediation is a mediated query. To retrieve the data from the disparate information sources, the mediated query is transformed into a query execution plan, which is optimized, taking into account the topology of the network of sources and their capabilities. The plan is then executed to retrieve the data from the various sources; results are composed as a message, and sent to the receiver.

Domain model: The domain model defines the different elements that are needed to implement the strategy in a given application: The domain model a collection of rich types (*semantic types, attributes, etc.*) and relationships (*is-a relationship*) defining the domain of discourse for the integration strategy. This declarative knowledge about the domain ontology is represented independent of the various information sources and represents the generic concepts associated with the domain under consideration. Semantic types resemble the different entities in the underlying domain. For example *Account, Person* can be entities in a financial domain. The attributes represents the generic features those semantic types can have. i.e. *bankBalance, creationDate* attributes of *Account* semantic type. Further, attributes can be used to infer relationships between different entities. For example the *holder* attribute of an *Account* could refer to a *person* semantic type. In some instance the attribute can constitute a basic type; either a string or a numeric value represented by the super semantic type, *basic*.

Context: Context axioms are used to capture different semantic, contextual, and ontological representations that the underlying data sources contain. The context definitions define the different interpretations of the semantic objects from the different sources' or receiver's point of view. We use a special type of attributes, *modifiers*, to define the context of a data type. For example *currency* modifier may define the context of objects of semantic type *moneyAmount,* when they are instantiated in a specific context (i.e., currency is USD in that specific context).

Sources: All sources are represented using the *Source* concept where the type of the sources could be any data source ranging from a relational table, an XML stream, to a web page. Different wrapper implementations, including the web data extraction engine *Cameleon* [12], provide different interfacing mechanisms to diverse sources.

Elevation Axioms: The sources and the domain model needs to be linked in order to facilitate mediation. This is achieved through the definition of *Elevation axioms*. Its usage is two fold. First, each source is given a Context definition. Second, each attribute of the source is elevated to a particular semantic object (instances of semantic types) that is represented in the Domain Model. This facilitates in bridging the link between the context-independent, *'generic'* domain model and the context dependent sources.

Conversion library: Finally, there is a conversion library, which provides conversion functions for each *modifier* that specifies the resolution of potential conflicts. The relevant conversion functions are gathered and composed during mediation to resolve the conflicts. No global or exhaustive pair-wise definition of the conflict resolution procedures is needed. The mediation is performed by a procedure, which infers from the query and the knowledge base a reformulation of the initial query in the terms of the component sources. The procedure itself is inspired by the abductive logic programming framework [27]. One of the main advantages of the abductive logic-programming framework is the simplicity in which it can be used to formally combine and to implement features of query processing, semantic query optimization, and constraint programming.

In the next section we would show how these concepts are applied in our case study.

4   Case Study

4.1   Electronic Bill Presentment and Payment Domain

In order to demonstrate the usage of the COIN framework, a subset of the standards, namely the '*Electronic Bill Presentment any Payment – (EBPP)*' domain is selected. The EBPP domain is a rich subset of the financial services messaging frameworks that have considerable amount of heterogeneities. The main standards are OFX, IFX for intra-bank payment schemes and SWIFT for inter-bank payment and funds transfer.

The overall functionality can be visualized from Figure 1 in using those standards. The focus is on the analysis of various heterogeneities that lie among these standards and financial systems as well as how they are handled using COIN. The key intermediaries in an EBPP scheme are as follows:

- Biller Payment Provider (BPP) is an agent (usually a financial institution) of the Biller that originates and accepts payments on behalf of the Biller.
- Biller Service Provider (BSP) is an agent of the Biller that provides an electronic bill presentment and payment service for the Biller
- Customer Payment Provider (CPP) is an agent (usually a financial institution) of the Customer that originates payments on behalf of the Customer.
- Customer Service Provider (CSP) is an agent of the Customer that provides an interface directly to customers, businesses, or others for bill presentment. A CSP enrolls customers, enables presentment, and provides customer care, among other functions.
- Financial Institution (FI) is an organization that provides branded financial services to customers. Financial Institutions develop and market financial services to individual and small business customers. They may serve as the processor for their own services or may choose to outsource processing.

Both IFX and OFX provide XML based messaging framework for individuals as well as businesses in bill payment and presentment electronically. But the most acclaimed inter-bank fund transfer framework, SWIFT uses a non XML messaging protocol and recently went through a major restructuring in phasing out one of the most utilized messaged for inter-bank customer fund transfer, the M100, and introduced modified versions of MT103 and MT103+.

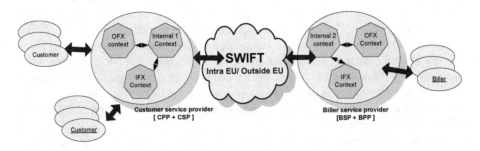

Fig. 1. Interfaces in EBPP

In order to depict the usage of COIN in EBPP mediation in a practical scenario, we have broken down the analysis to three main areas that spans from a customer initiating a Bill payment to its subsequent verification by the Biller. In addition to analyzing these standards separately we address how they are utilized in practical scenarios. All three standards inevitably require interfacing with the internal accounting and financial system of a financial system to make a successful payment from the customer to the Biller as in figure 1. For example a bill payment from a customer might interact with an FI's IFX based system which in turn has to interface with customer's bank's internal accounting system. Then to facilitate the inter-bank funds transfer to the Biller's bank, a separate interfacing is required with a global inter bank messaging framework like SWIFT. Finally at the Biller's bank it needs to be represented and stored in its proprietary financial system. Finally the biller should be able to view the payments through its bank's bill presentment system which possibly utilizes OFX standard, where the internal representation needs to be transformed according to OFX's syntax and semantics. Therefore we have introduced a hypothetical internal system that represents the true nature of a realistic situation to bridge the gap between the financial standards as modeled in a real-life situation. The conflicts analysis and mediation with the diverse financial standards have been analyzed with respect to the hypothetical internal system of a Financial Institution which could be an in-house developed system or third-party (off the shelf) system. This internal system is represented by the term 'internal context'. Following are the three main areas analyzed in the case study.

- Mediation between an internal context and OFX context.
- Mediation between an internal context and IFX context.
- Mediation between an internal context and SWIFT context.

The IFX, OFX and SWIFT contexts represent the semantics and definitions adopted by IFX, OFX and SWIFT messaging frameworks respectively. SWIFT distinguishes intra European Union (EU) fund transfer and outside EU fund transfers for accounting for inter-bank charges.

Figure 2 represents the context independent, COIN domain ontology for the EBPP domain denoting some of the key the concepts used by IFX, OFX, SWIFT and financial institution's own internal schema. This was constructed by exploring the business domain in EBPP and the relevant message handling semantics used in these diverse standards. The semantic types (entities) represents the business entities that encompass the main functionalities in EBPP Industry The sources and their conflict are mapped to these semantic types (entities). The semantic types denote the entities and their relationships in the EBPP domain like *Payment*, *Account* etc. *is-a* relation denotes an inheritance relationship between semantic types. A semantic type may have certain *Attributes* (e.g., payment has payee, payer accounts, amount etc). The entities that constitute conflicts in these contexts are modeled through *modifiers*. As an example, the *paymentAmount* can include/exclude various taxes in different contextual representations and in SWIFT it would incur an additional inter-bank service charge. These are represented by COIN modifiers *paymentScheme*, *includesInterBankCharge* respectively. Further, monetary amounts could be expressed in different currencies. This is modeled using the *currency* modifier for the super-semantic type *moneyAmount*. This represents how COIN models inheritance of contextual knowledge for different entities.

In an actual scenario, the heterogeneities of the standards and the mappings needed for mediation would be analyzed and formulated by a business analyst or a person working for the respective Financial Institution. The following sections addresses each of these three cases separately

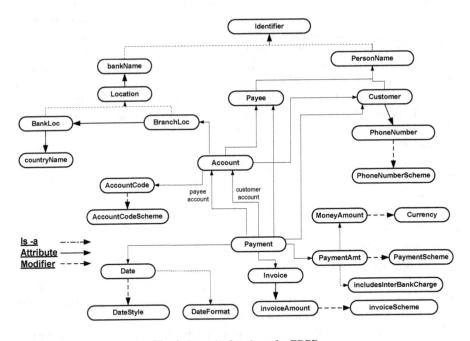

Fig. 2. Domain Ontology for EBPP

4.2 Internal Schema Versus OFX

First we will look at the mediations between OFX and an internal schema of a financial institution. Table 2 summarizes the heterogeneities identified in the two schemas. As denoted in COIN's mediation strategy, the modifiers and relevant conversion functions are the main ingredients in facilitating the mediation for a particular heterogeneity exiting between two different contexts. As shown in the table, there are different types of heterogeneities between the two contexts. The significant conflicts are payment amount, currency type and Account code reference identifiers. They are discussed below.

Table 2. Conflicts in Internal and OFX Contexts

Conflict	Internal Context	OFX Context	Mapped modifier (refer Fig 2)
Payment amount	Net amount without tax	Net + tax amount	*PaymentScheme*
Account Location Identifier – BANK reference	Bank identifier represented in the internal scheme	Bank Identifier depends on the Bank's country of Incorporation.	*BankLoc*
Account Location Identifier – BANK BRANCH reference	Branch identifier of the account	Branch Identifier dependent on the bank's country of incorporation.	*BranchLoc*
Payment due date format	European format	US format	*DateFormat*
Payment due date Style	dd/mm/yyyy 03/03/2003	Yyyymmdd 20030303	*DateStyle*
Account type code	CHECKING,SAVINGS etc	CHK,SVG etc	*AccountCodeScheme*
Currency type (Exchange rate)	"EUR"	Currency of country of incorporation of payee bank	*currency*
Phone number format	415.445.4345	1-415-445-4345	*PhoneNumberScheme*

Payment Amount – The mediation strategy for payment amount is as follows. The mediator needs to apply two conversion functions in order to obtain the mediated payment amount, namely the currency conversion inherited from the *moneyAmount* super semantic type, and the tax adjustment for the payment. For simplicity let's assume that in both contexts the currency is denoted in three letter ISO 4217 format (i.e. USD, GBR, and EUR etc).

Assume that the query *'select AMOUNT FROM PAYMENT'* is called in *OFX* context;

First, payment amount is adjusted for the tax inclusion. For simplicity let's assume that the applicable tax is 'GST'. Then;

$$\text{Payment}_{OFX} = (\text{payment}_{INTERNAL} + \sum \text{GST amount for payment}_{OFX} * \text{payment}_{INTERNAL}) \quad (1)$$
$$* \text{ Exchange Rate ("EUR", OFF\_CUR,DATE\_OF\_TRANSACTION)}$$

In the COIN framework, the mediation formulas are translated into logical expressions of the COIN theoretical model [1]. Later these expressions are implemented in prolog and evaluated by an abduction engine implemented in the same language [13]. The following describes the logical representation of the formula (1) for this example.

The formula below describes a non-commutative mediation of *paymentType* object depending on its modifier *paymentScheme*, which in this case holds the values "*noTax*" or "*withTax*". The *Ctxt* defines the destination context. The conversion in simple terms would be to retrieve the Rate for the tax "GST" from the elevated relation '*OFX_TAX_TYPES_p*' which is an elevation mapped to relation '*OFX_TAX_TYPES*' under *OFX* Context (The destination context in this case) and utilizes in the tax calculation. The *value* predicate in the formula defines a value of a particular semantic object under a certain context.

```
cvt(noncommutative,paymentAmt,_O,paymentScheme,Ctxt,"nota
x",Vs,"withtax",Vt) ⇐
       value(TaxName,Ctxt,"GST"),
       'OFX_TAX_TYPES_p'(TaxName,_,Rate),
       value(Rate,Ctxt,RR),
       (Vtemp is RR * Vs),
       (Vt is Vs + Vtemp).
```

Further, this resembles an *Equational ontological heterogeneity* addressed in [5], which is a clear example of differences in the contexts of OFX and internal contexts. But the ontological conflict has been transformed into a contextual heterogeneity by way of matching the definitional equations as in [5].

Then, this tax adjusted payment needs to be mediated to the currency of *OFX* context. This requires a *dynamic modifier* to extract the currency value depending on the official currency in the incorporated country of the payee's bank as given below.

$$OFF\_CUR_{OFX} = Currency_{OFX}(payment) \quad \Leftarrow \quad AID = Payee\ Account\ of\ Payment_{INTERNAL}$$
$$BRANCH_{OFX} \quad \Leftarrow \quad Branch\ of\ Account\ AID_{OFX}$$
$$BANK_{OFX} \quad \Leftarrow \quad Bank\ of\ BRANCH_{OFX} \tag{2}$$
$$COUNTRY_{OFX} \quad \Leftarrow \quad country\ of\ Incorporation\ of\ BANK_{OFX}$$
$$OFF\_CUR_{OFX} \quad \Leftarrow \quad official\ currency\ of\ COUNTRY_{OFX}$$

The following logical representation describes how the value of modifier *currency* for *paymentAmount* is obtained for *OFX* context dynamically through the relationships between semantic objects.

```
modifier(paymentAmt,_O,currency,ofx,M) ⇐
( attr(_O,paymentRef,Payment),
attr(Payment,payeeAct,Account),
              attr(Account,location,Location),
              attr(Location,bank,Bank),
              attr(Bank,countryIncorporated,Country),
              attr(Country,officialCurrency,M))).
```

For example the predicate *attr (Payment,payeeAct,Account)* defines the attribute relationship 'payeeAct' between the *Payment* and *Account* semantic objects. This relation can be mapped to underlying relationships in different contexts as shown in the following logical representation.

```
attr(Payment,payeeAct,PayeeAcct) ⇐

   ('INTERNAL_PAYMENT_p'(Payment,_,_,_,_,_,PayeeAcct,_).

attr(Payment,payeeAct,PayeeAcct) ⇐

   ('OFX_PAYMENT_p'(Payment,_,_,_,_,_,PayeeAcct,_).
```

The two statements correspond to how the attribute relation *payeeAcct* has been elevated to two elevation relations with their attributes, mapped in *INTERNAL* and *OFX* contexts.

Account Type Code – This is represented as heterogeneity in enumerated data types in defining the account type codes in the three contexts. The following summarizes the enumerated data mapping in the three contexts. Since there can be more than two types of financial standards, rather than having mappings between each standard , we adopt a 'Indirect conversion with ontology inference' strategy [13] where we represent the different account types in the ontology itself and providing mapping between the context independent ontology's enumerated type and the context sensitive type codes. The context model would then map each security type context construct into its corresponding security type ontology construct.

Therefore the conversion from *INTERNAL* to *OFX* would be,

$Account\_type_{OFX}$

$$(Account\_type_{INTERNAL}('CHK')) \Longleftarrow \begin{array}{ll} ONTOLOGY\_TYPE_{INTERNAL} & = 'CHKA' \text{ [table INTERNAL]} \\ ONTOLOGY\_TYPE_{NONE} & = 'CHKA' \text{ [table Ontology]} \\ ONTOLOGY\_TYPE_{OFX} & = 'CHKA' \text{ [table OFX]} \\ OWN\_TYPE \ ('CHK')_{OFX} & = 'CHECKING' \text{ [table OFX]} \end{array} \quad (3)$$

Ontology : Account types Table Ontology		Mapping between Internal and Ontology - Table INTERNAL	
ONTOLOGY_TYPE	Description	ONTOLOGY_TYPE	OWN_TYPE
CHKA	Checking account	CHKA	CHK
SVGA	Savings account	SVGA	SVG
MNYMRTA	Money Market Account	MNYMRTA	MNYMRT
CRLINEA	Credit Line Account	CRLINEA	CRLINE

Mapping between OFX and Ontolo - Table OFX		Mapping between IFX and Ontology - Table IFX	
ONTOLOGY_TYPE	OWN_TYPE	ONTOLOGY_TYPE	OWN_TYPE
CHKA	CHECKING	CHKA	DDA
SVGA	SAVINGS	SVGA	SDA
MNYMRTA	MONEYMRKT	MNYMRTA	MMA
CRLINEA	CREDITLINE	CRLINEA	CDA

4.3 Internal Schema Versus IFX

After looking at some of the interoperability issues between internal context and OFX, now we would delve into the newer standard, IFX, which has more features and detailed representations. Table 3 shows the different types of heterogeneities. The conflicts of *account type*, *date format*, *phone number format* and *currency types* are similar to the OFX scenarios. The new conflicts are the extended conflicts identified in *payment amount* and introduction of *invoice* related conflicts.

Both IFX and OFX handle complex business payment transactions for business customers. This requires incorporating multiple invoice details attached to the payment aggregates when both the biller and customer are business entities. The older OFX provides a basic mechanism of incorporating invoice details like invoice discounts, line items in invoices etc. But the newer IFX extends this by providing more elaborate aggregates constituting different tax schemes as well as fees (late fees, FoRex fees, etc.) that are applicable to invoice.

Mediating Invoice Amount

Each payment can have at least one invoice aggregate that represent the different invoices paid through a particular invoice. In an internal schema the invoice amount might be represented as the net amount, where the taxes and fees would be aggregated when the bill is presented or invoiced. But the IFX context, the Invoice amount consists of the various taxes and fees that could be added to the net amount.

Table 3. Conflict between Internal and IFX contexts

Conflict	Internal Context	IFX Context	Mapped modifier (Refer Fig 2)
Payment amount	Net amount	$Net + \sum tax\ amount + \sum Fees$	PaymentScheme
Payment due date format	European format	US format	DateFormat
Payment due date Style	dd/mm/yyyy 03/03/2003	Yyyy-mm-dd 2003-03-03	DateStyle
Account type code	SVG,MNYMRT,CRLINE, CHK etc	SDA,MMA,CCA,DDA etc	AccountCodeScheme
Invoice Amount	Net amount	$Net + \sum tax\ amount + \sum Fees$	InvoicePayment-Scheme
Currency type (Exchange rate)	"GBP"	Currency of country of incorporation of payee bank	Currency
Phone number format	415.445.4345	1-415-4454345	PhoneNumberScheme

The mediation between the two invoice amounts represents an equational ontological conflict (EOC) [5] that would be resolved through introduction of a set of modifiers that would match the two different definitional equations. Each invoice would have multiple fees .i.e. an invoice would have FoRex, late payment fees, import fees as well as multiple taxes like GST, withholding taxes etc

Therefore the relationship between the two definitional equations for invoice amount is:

$$InvoiceAmount_{IFX} = InvoiceAmount_{internal}$$
$$+ \sum (InvoiceAmount_{internal} * FeeRate_{IFX}) + \sum (FixedFee_{IFX}) \qquad (4)$$
$$+ \sum (InvoiceAmount_{internal} * TaxRate_{IFX}) + \sum (FixedTax_{IFX})$$

Let us say we executed the query `select INVOICE_AMOUNT from INTERNAL_INVOICE` in *IFX* context where the relation `INTERNAL_INVOICE` is defined for *internal* context.

The following shows the mediated SQL query automatically generated by the COIN mediation framework considering all the conflicts associated between *internal* and *IFX* contexts:

```
select
(internal_invoice.INVOICE_AMOUNT+(((internal_invoice.INVOICE_AMOUNT*ifx_
tax_types.AMOUNT)+(internal_invoice.INVOICE_AMOUNT*ifx_tax_types2.AMOUNT
))+(ifx_fees_types.AMOUNT+(internal_invoice.INVOICE_AMOUNT*ifx_fees_type
s2.AMOUNT))))
    from (select 'GST', TYPE, AMOUNT from   ifx_tax_types
        where  TAX_NAME='GST') ifx_tax_types,
        (select 'IMPORT', TYPE, AMOUNT from   ifx_tax_types
        where  TAX_NAME='IMPORT') ifx_tax_types2,
        (select 'LATE', TYPE, AMOUNT from   ifx_fees_types
        where  FEES_NAME='LATE') ifx_fees_types,
        (select 'DELIVERY', INVOICE_NO from   ifx_invoice_fees
        where  FEE_NAME='DELIVERY') ifx_invoice_fees,
        (select 'LATE', INVOICE_NO from   ifx_invoice_fees
        where  FEE_NAME='LATE') ifx_invoice_fees2,
        (select 'IMPORT', INVOICE_NO from   ifx_invoice_taxes
        where  TAX_NAME='IMPORT') ifx_invoice_taxes,
        (select 'GST', INVOICE_NO from   ifx_invoice_taxes
        where  TAX_NAME='GST') ifx_invoice_taxes2,
        select   INVOICE_NO,    PAYMENT_ID,    INVOICE_AMOUNT,    DESCR,
        INVOICE_DATE,
        DISCOUNT_RATE,DISCOUNT_DESC   from   internal_invoice)   internal_
        invoice,
        (select 'DELIVERY', TYPE, AMOUNT rom   ifx_fees_types
        where  FEES_NAME='DELIVERY') ifx_fees_types2
            where ifx_invoice_fees.INVOICE_NO = ifx_invoice_fees2.INVOICE_NO
            and ifx_invoice_fees2.INVOICE_NO = ifx_invoice_taxes.INVOICE_NO
            and ifx_invoice_taxes.INVOICE_NO = ifx_invoice_taxes2.INVOICE_NO
            and ifx_invoice_taxes2.INVOICE_NO = internal_invoice.INVOICE_NO
```

Some readers may have so far considered that identifying and resolving semantic heterogeneity is a small matter of handling date formats, currency exchange, and other accounting conventions. We observe now that the net effect and accumulation of such small matters makes the programmer's task impossible. A programmer not equipped with the COIN mediation system must devise and create the above query. A programmer using the COIN mediation system can type the original query: '*select INVOICE_AMOUNT from INTERNAL_INVOICE*' in *IFX* context and rely on COIN to automatically mediate the query. The application gains in clarity of design and code, as well as in scalability. The sharing of domain knowledge, context descriptions, and conversion functions improve the knowledge independence of the programs and their maintainability.

4.4   Some Insight to Conflicts Analysis Between Internal and SWIFT Contexts

The SWIFT protocol is mainly involved in inter-bank cross border transactions. It uses globally unique identifiers for bank code like BIC, BEI. For e.g. the BCI code comprise of concatenation of bank code, country code and location code (defined by ISO 9362), compared to just a bank code representation used in internal schema. This peculiar heterogeneity requires a non-commutative building up of a composite bank identifier when mediating from *internal* to *SWIFT* context. The following represents a logical formula for the mediation for the concatenation. The predicate notations were discussed in a previous example.

```
cvt(noncommutative,bankLoc,O,idType,Ctxt,"single",Vs,"composite",Vt) ⇐
('SWIFT_BANK_BCI_p'(BANK, LOC, COUNTRY), value(BANK,Ctxt,Vs),
value(LOC,Ctxt,Locc), value(COUNTRY,Ctxt,Countryc),
(Vtemp is Vs + Locc), (Vt is Vtemp + Countryc))).
```

Usage of Sub Contexts

Under the *SWIFT* context, depending on whether the transaction is between financial institutions inside the EU or outside, a bank handling fee is credited to the payment amount. This can be modeled using the *sub context* concept of COIN. A sub context derives all the super context based modifier values while having specialized modifier values for extended features. The following logical formulas denote how this can be modeled in COIN

```
is_a(swift_intraEU,swift)
is_a(swift_outsideEU,swift)
```

Then a query like '*select amount from payment*' in *outsideEU* context, called on a relation defined for internal context, is resolved by adding the handling charges on top of the local applicable tax (inherited from *SWIFT* context) as denoted in the following mediated datalog.

```
answer('V15'):-
'INTERNAL_PAYMENT'('V14', 'V13', 'V12', 'V11', 'V10', 'V9', 'V8', 'V7'),
'TAX_TYPES'("GST", 'V6', 'V5'), 'V4' is 'V5' * 'V12',
'V3' is 'V12' + 'SWIFT_CHARGE_TYPES'
("outsideEU", V2', 'V1'),
'V15' is 'V1' + 'V3'.
```

Note that although datalog and prolog representations are used internally within COIN and shown in this paper, the actual COIN system provides a graphical and user-friendly interface so that data administrators setting up the knowledge representations (e.g., domain models, context)   need not know anything about these internal representations.

5   Conclusion and Future Work

We identified different semantic, ontological heterogeneities that exist in different financial messaging standards. It showed that indeed mediation between these is not a trivial task, yet is critical and important to the globalization of the financial industry. Further we show that an effective answer is to have a mediation service that provides automatic and a transparent mediation without requiring engineering new standards.

Table 4. Temporal Heterogeneities

	Internal Schema	OFX	IFX	SWIFT 03/103+
Price	Net	Net + tax of 5% on and before 2000 , Net + tax of 2% after 2000	Net + tax of 5% on and before 2000 , Net + tax of 2% after 2000	(Net + tax of 5% on and before 2000 , Net + tax of 2% after 2000) + inter-bank charges.
Currency	FFR on and before 2000, EUR after 2000.	Currency of country of incorporation of payee bank	Currency of country of incorporation of payee bank	Explicitly mentioned- ISO 4217

We have shown that the COIN approach is capable of mediating the different heterogeneities that exist in different financial standards and internal contexts of Financial Institutions. Our approach in modeling a business domain and mapping different contextual representations and values through a declarative manner demonstrates the extensibility, flexibility and user-friendliness of the COIN framework.

One aspect that is lacking in COIN and that we are currently investigating is the modeling temporal heterogeneities like the examples denoted in table 4. We are currently studying different aspects of temporal heterogeneities which are sources of conflicts among financial standards.

Acknowledgements

The authors wish to acknowledge the extensive help of Aykut Firat, Hongwei Zhu, Philip Lee and Allen Moulton of MIT. The research reported herein has been supported, in part, by the Singapore-MIT Alliance.

References

[1] A.Firat ."Information Integration using Contextual Knowledge and Ontology Merging", PhD Thesis, MIT,2003

[2] C.H. Goh, S.Bressan.S.Madnick,M.Siegel, "Context Interchange :New Features and Formalisms for the Intelligent Integration of Information", ACM TOIS, vol. 17,pp 270-293,1999.

[3] A.Bressan , C.H. Goh, "Answering Queries In Context", Proceedings of "Flexible Query Answering Systems". Third International Conference, FQAS, 1998, Roskild,Denmark.

[4] S.Madnick,A.Moulton,M.Siegel, "Semantic Interoperability in the Fixed Income Securities Industry: A Knowledge Representation Architecture for dynamic integration of Web-based information", HICSS,Hawai,2003

[5] S.Madnick, A.Firat, B.Grosof, "Knowledge Integration to overcome Ontological Heterogeneity: Challenges from Financial Information Systems",pp. 183-194,ICIS,Barcelona,Spain, 2002

[6] S.Madnick,A.Moulton,M.Siegel, "Context Interchange Mediation for Semantic Interoperability and Dynamic Integration of Autonomous Information Sources in the Fixed Income Securities Industry", (WITS), Barcelona, Spain, December 14-15, 2002, pp.61-66

[7] S.Madnick,S. Bressan, C.H. Goh, T. Lee, and M. Siegel "A Procedure for Mediation of Queries to Sources in Disparate Context", Proceedings of the International Logic Programming Symposium, October 1997

[8] S.Madnick, S. Bressan, C. Goh, N. Levina, A. Shah, M. Siegel ,"Context Knowledge Representation and Reasoning in the Context Interchange System" , *Applied Intelligence: The International Journal of Artificial Intelligence, Neutral Networks, and Complex Problem-Solving Technologies,* Vol 12, Number 2, September 2000, pp. 165-179

[9] Open Financial Exchange Specification OFX 2.0.2, Open Financial Exchange, http://www.ofx.net/ofx/de_spec.asp

[10] Interactive Financial Exchange –IFX version 1.5, IFX Forum, Inc,
      http://www.ifxforum.org/ifxforum.org/standards/standard.cfm

[11] Society for Worldwide Interbank Financial Telecommunication (S.W.I.F.T), Standard
      Release 2003, http://www.swift.com/index.cfm?item_id=5029

[12] S.Madnick, A. Firat and M. Siegel, "The Caméléon Web Wrapper Engine", *Proceedings
      of the VLDB2000 Workshop on Technologies for E-Services,* September 14-15, 2000

[13] S.Madnick, A. Moulton and M. Siegel "Semantic Interoperability in the Securities
      Industry: Context Interchange Mediation of Semantic Differences in Enumerated Data
      Types", *Proceedings of the Second International Workshop on Electronic Business
      Hubs: XML, Metadata, Ontologies, and Business Knowledge on the Web* (WEBH2002),
      Aix En Provence, France, September 6, 2002

[14] Kuhn, E., Puntigam, F., Elmagarmid A. (1991). Multidatabase Transaction and Query
      Processing in Logic, Database Transaction Models for Advanced Applications, Morgan
      Kaufmann Publishers.

[15] Litwin, W., Abdellatif, A. (1987), "An overview of the multi-database manipulation
      language MDSL". Proceedings of the IEEE, 75(5):621-632.

[16] Goh, C. H. (1997), "Representing and Reasoning about Semantic Conflicts in
      Heterogeneous Information Systems, MIT Ph.D. Thesis.

[17] Arens, Y., Knoblock, C., Shen, W. (1996). Query Reformulation for Dynamic
      Information Integration. Journal of Intelligent Information Systems 6(2/3): 99-130.

[18] Batini, C., Lenzerini, M., Navathe, S. B. (1986) "A Comparative Analysis of
      Methodologies for Database Schema Integration", ACM Computing Surveys 18(4):
      323-364.

[19] Landers, T., Rosenberg, R (1982) "An Overview of MULTIBASE", International
      Symposium on Distributed Data Bases", 153-184

[20] Breitbart, Y., Tieman.L. (1984), "ADDS - Heterogeneous Distributed Database System",
      Proceedings of the Third International Seminar on Distributed Data Sharing Systems, 7- 24.

[21] Scheuermann, P., Elmagarmid, A. K., Garcia-Molina, H., Manola, F., McLeod,
      D.,Rosenthal, A., Templeton, M. (1990), "Report on the Workshop on Heterogeneous
      Database Systems" held at Northwestern University, Evanston, Illinois, December 11-13,

[22] Ahmed, R., De Smedt, P., Du, W., Kent, W., Ketabchi, M., Litwin, W.,, Rafii,A.,,Shan,
      M. (1991)." The Pegasus Heterogeneous Multidatabase System". IEEE Computer
      24(12): 19-27.

[23] Collet, C., Huhns, M. N., Shen, W. (1991), "Resource Integration using a large
      knowledge base in Carnot", IEE Computer, 24(12):55-63.

[24] Kuhn, E., Ludwig, T. (1988), "VIP-MDBS: a logic multidatabase system", Proceedings
      of the first international symposium on Databases in parallel and distributed systems,
      p.190-201, December 05-07, Austin, Texas, USA.

[25] Litwin, W. (1992). "O*SQL: A language for object oriented multidatabase
      interoperability". In Proceedings of the Conference on IFIP WG2.6 Database Semantics
      and Interoperable Database Systems (DE-5) (Lorne, Victoria, Australia), D. K. Hsiao, E.
      J. Neuhold, and R. Sacks-Davis, Eds. North-Holland Publishing Co., Amsterdam, The
      Netherlands, 119-138.

[26] Baral, C., Gelfond, M. (1994). "Logic Programming and Knowledge Representation",
      Journal of Logic Programming, 19,20:73-148.

[27] Kakas, A. C., Michael, A. (1995). "Integrating abductive and constraint logic
      programming", To appear in Proc. International Logic Programming Conference.

[28] S. Patel and A. Sheth, "Planning and Optimizing Semantic Information Requests using Domain Modeling and Resource Characteristics," Proceedings of the 6th Intl Conf on Cooperative Information Systems (CoopIS), Trento, Italy, September 5-7, 2001, pp. 135-149.

[29] P. Ziegler and K R. Dittrich. (2004)., "User-Specific Semantic Integration of Heterogeneous Data: The SIRUP Approach". International Conference on Semantics of a Networked World (IC-SNW 2004), Paris, France, June 17-19, 2004

[30] Wache, H., Vogele, T., Visser, U., Stuckenschmidt, H., Schuster, G., Neumann, H., Hubner, S., (2001), "Ontology-Based Integration of Information – A Survey of Existing Approaches", Proceedings of the IJCAI-01 Workshop on Ontologies and Information Sharing, Seattle, USA, 4 –5 August, 2001.

[31] M. Lenzerini. (2002). "Data integration: a theoretical perspective", Proceedings of the twenty-first ACM SIGMOD-SIGACT-SIGART symposium on Principles of database systems, pp233 - 246

Author Index

Lecture Notes in Computer Science

For information about Vols. 1–3323

please contact your bookseller or Springer